**Green Energy and Technology**

More information about this series at http://www.springer.com/series/8059

Magno Trindade
Editor

# Increased Biodiesel Efficiency

## Alternatives for Production, Stabilization, Characterization and Use of Coproduct

Springer

*Editor*
Magno Trindade
Faculty of Exact Sciences and Technology
Federal University of Grande Dourados
Dourados, Mato Grosso do Sul
Brazil

ISSN 1865-3529 ISSN 1865-3537 (electronic)
Green Energy and Technology
ISBN 978-3-319-73551-1 ISBN 978-3-319-73552-8 (eBook)
https://doi.org/10.1007/978-3-319-73552-8

Library of Congress Control Number: 2017963015



Printed on acid-free paper

This Springer imprint is published by the registered company Springer International Publishing AG part of Springer Nature
The registered company address is: Gewerbestrasse 11, 6330 Cham, Switzerland

# Presentation

Despite countless valuable scientific texts and books highlighting the research into biofuels, the field is not fully explored in order to address all alternative sources of biomass, new technologies for biodiesel production and its characterization as well as the current green technologies on sample preparation for elemental determination and smart use of coproducts. In this book, the authors start to address the use of alternative non-food feedstock oilseed species to produce biodiesel, covering the new opportunities and challenges about the viability for exploring the biodiesel production in an economical way. New technologies for the synthesis of biodiesel via alternative protocols accompanied by alternative techniques for characterization are reported. The following chapters also address the current technologies to cover the alternative blends composed by multifunctional antioxidants (sometimes called secondary antioxidants) which are followed by the uses of chalcones and their derivatives to overcome the drawback associated with the biodiesel quality and provide its stabilization more efficiently and more cost-effectively. Moreover, efforts were made to cover some aspects based on biodiesel characterization, providing alternatives to overcome some of the limitation with the uses of traditional method, especially, showing the potential use of portable systems and the fluorescence spectroscopy as an alternative analytical tool for, in situ and in real time, monitoring the oxidative stability of biodiesel. The concerns about the recent advances on green technologies for sample preparation and for determination of the content of metallic species are also addressed, focusing in the alternatives based on nondestructive measurement for elemental determination in the entire biodiesel production chain. The last chapter brings the current technologies for smart use of coproducts, addressing how the biodiesel byproduct can provide alternative uses for glycerol as a source of energy and high valuable chemicals. Finally, in the book, we focused in many topics—sometimes neglected in previously published books. This allows the readers to understand the challenges regarding technological resources and research required to overcome the drawback associated with the biodiesel production chain, starting from the choice of non-food feedstocks up to the use of coproducts.

# Contents

# Abbreviations and Symbols

| | |
|---|---|
| %SYN | Percent of synergism |
| [Hmim][PF6] | 1-Hexyl-3-methylimidazolium hexafluorophosphate |
| [Hmim][TF2N] | 1-Hexyl-3-methylimidazolium bis(trifluoromethylsulfonyl) imide |
| [Omim][TF2N] | 1-Octyl-3-methylimidazolium bis(trifluoromethylsulfonyl) imide |
| AAS | Atomic absorption spectrometry |
| AEM-DGFC | Anion-exchange membrane-direct glycerol fuel cell |
| AES | Atomic emission spectrometry |
| AFS | Atomic fluorescence spectrometry |
| AgNPs | Silver nanoparticles |
| ALZ | Alizarin |
| ANP | National Agency of Petroleum, Natural Gas and Biofuels |
| ASTM | American Society for Testing and Materials |
| AuNPs | Gold Nanoparticles |
| B100 | Pure biodiesel |
| B20 | 20% Blend of biodiesel in diesel |
| BEQ | 1,4-Bis(ethylamino)-9,10-anthraquinone |
| BHA | Butylhydroxyanisole or butylated hydroxyanisole |
| BHT | Butylhydroxytoluene or butylated hydroxytoluene |
| CA | Citric acid |
| CCD | Central composite design |
| CFA | Caffeic acid |
| CG | Gas chromatography |
| CHI | Chalcone isomerase |
| CICP | Cetylpyridinium chloride |
| CoA | Coenzyme A |
| CPE | Cloud point extraction |
| CRM | Certified reference materials |
| CSH | Chalcone synthase |

| | |
|---|---|
| CTAB | Hexadecyltrimethylammonium bromide |
| DAFC | Direct alcohol fuel cells |
| DDBS | Sodium dodecyl benzene sulfonate |
| DGFC | Direct glycerol fuel cell |
| DHA | Dihydroxyacetone |
| DHQ | 1,4-Dihydroxanthraquinone |
| DLLME | Dispersive liquid–liquid microextraction |
| DME | Bio-dimethyl ether |
| DMF | N,N-dimethyl formamide |
| DSC | Differential scanning calorimetry |
| DTBP | 2,6-Di-tert-butylphenol |
| DTPA | Diethylenetriaminepentaacetic acid |
| EDTA | Ethylenediaminetetraacetic acid |
| EF | Enrichment Factor |
| EN | European Norms |
| EPE | Energy Research Company |
| ER | Extraction recovery |
| ET AAS | Electrothermal atomic absorption spectrometry |
| EtOH | Bioethanol |
| EU | European Union |
| F AAS | Flame atomic absorption spectrometry |
| F AES | Flame atomic emission spectrometry |
| FA | Ferulic acid |
| FAs | Fatty acid |
| FAME | Fatty acid methyl esters |
| FS FAAS | Fast sequential flame atomic absorption spectrometry |
| FSB | Fried soybean oil |
| FT | Fischer–Tropsch fuels |
| FTIR | Fourier transform Infrared |
| GA | Gallic acid |
| GC | Gas chromatography |
| GC-FID | Gas chromatography with flame ionization detector |
| GC-MS | Gas chromatography coupled to mass spectrometry |
| GEOR | Glycerol electrooxidation reaction |
| GF AAS | Graphite furnace atomic absorption spectroscopy |
| GHG | Greenhouse gas |
| GT-C | Green tea chloroform extract |
| GT-M | Methanol extracts of green tea |
| HF-LPME | Hollow fiber liquid-phase microextraction |
| HLB | Hydrophilic–lipophilic balance |
| HLLE | Homogeneous liquid–liquid extraction |
| HPLC | High-performance liquid chromatography |
| HVO | Hydrotreated vegetable oil |
| ICP-MS | Inductively coupled plasma mass spectrometry |
| ICP-OES | Inductively coupled plasma optical emission spectrometry |

| | |
|---|---|
| IL-DLLME | Ionic liquid dispersive liquid–liquid microextraction |
| ILs | Ionic liquids |
| IP | Induction period |
| IRAM | Argentine Institute for Standardization and Certification |
| LLE | Liquid–liquid extraction |
| LLLME | Liquid–liquid–liquid microextraction |
| LPME | Liquid-phase microextraction |
| LPME-SFO | Liquid-phase microextraction based on solidified of floating organic drop |
| MEA | Membrane exchange assembly |
| MEMR | Indonesian Ministry of Energy and Mineral Resources |
| MeOH | Biomethanol |
| MFC | Microfluidic fuel cell |
| MIP-OES | Microwave-induced plasma optical emission spectrometry |
| MOM | Methoxymethyl protecting groups |
| MWCNTs | Multi-walled carbon nanotubes |
| NPs | Nanoparticles |
| OCV | Open circuit voltage |
| OIT | Oxidation induction time |
| OOT | Oxidation onset temperature |
| OSR | Oilseed radish |
| OSRB | Oilseed radish biodiesel |
| PEM | Proton exchange membrane |
| PG | Propyl gallate or propyl 3,4,5-trihydroxybenzoate |
| PH-M | Pomegranate hull |
| PNPB | National Biodiesel Production and Usage Program |
| ppm | Parts per million |
| PROALCOOL | National Alcohol Program |
| PROBIODIESEL | Brazilian Program of Technological Development for Biodiesel |
| PRO-ÓLEO | Plan of Production of Vegetable Oils for Energy Purposes |
| p-TSA | p-Toluenesulfonic acid |
| PY | Pyrogallol or 1,2,3-trihydroxybenzene/pyrogallol |
| QC | Quercetin |
| RANP | Resolution of the National Oil Agency |
| Rf | Retention factor |
| RFS | Renewable fuel standard |
| RHE | Reversible hydrogen electrode |
| ROS | Reactive oxygen species |
| SAR | Structure–activity relationship |
| SB | Soybean biodiesel |
| SB-35 | Solvent blue 35 |
| SDME | Single-drop microextraction |

| | |
|---|---|
| SDS | Sodium dodecyl sulfate |
| SFME | Surfactant-free microemulsion |
| SOFC | Solid oxide fuel cell |
| STS | Stilbene synthase |
| TBHQ | *tert*-Butylhydroquinone or *tert*-butylated hydroxyquinone |
| TG | Thermogravimetry |
| THFA | Tetrahydrofurfuryl alcohol |
| TLC | Thin-layer chromatography |
| U.S. | United States |
| UA-SMDLLME-SFO | Ultrasonic-assisted supramolecular dispersive liquid–liquid microextraction based on solidification of a floating organic droplet |
| US EPA | United States Environmental Protection Agency |
| UV-Vis | Ultraviolet-visible |
| VA-LLME | Vortex-assisted liquid–liquid microextraction |
| WCAES | Tungsten coil atomic emission spectrometry |
| WCO | Waste cooking oil |
| WWII | World War II |
| $\theta_{Sb}$ | Antominy coverage degree |

# Chapter 1
# Renewable Energy Sources: A Sustainable Strategy for Biodiesel Productions

**Kássia R. N. da Silva, Marcela Z. Corazza and Jorge L. Raposo Jr.**

**Abstract** Nowadays, the world energetic scenario is moving toward a search for renewable energy matrices, allowing energy sources to be enlarged to produce diesel and/or biodiesel. Biodiesel is a renewable and clean-burning biofuel that is gradually being incorporated into conventional diesel as a means of limiting the production of fossil fuels to motivate the productivity of "ecologic friendly" biofuel. However, some oilseed species used as raw materials, even those with lower operational cost for the production of the vegetable oil, are still a great challenge for the petrochemical industry, given the economic competitiveness of biodiesel against the diesel obtained from fossil fuels. The use of alternative oilseed species to produce biodiesel may increase costs, which may exceed the price of the barrel of diesel, leading to non-viable biodiesel production. On the other hand, the high taxes on petroleum products can reduce the difference between them, allowing biodiesel production to become economically viable. This situation makes the exploration of economical alternative non-food feedstocks (e.g., *Jatropha curcas*) an important research topic. Biodiesel is made from diverse feedstocks including soybean, rapeseed/canola, sunflower, safflower, palm, *Jatropha curcas*, *Raphanus sativus*, cottonseed, corn, peanut, coconut, recycled cooking oil, animal fats, and also algae. In the context of renewable sources for biodiesel energy, the great biodiversity of alternative oilseeds and others oily biomass makes Brazil, the USA, Canada, and some countries in Europe and Asia greater potential markets to produce biofuels from these alternative sources. The incentives for using renewable species for biodiesel production rather than conventional petroleum diesel fuel are the absence of sulfur or aromatic compounds, high content of oil, low cost of crop maintenance,

K. R. N. da Silva · J. L. Raposo Jr. (✉)
Institute of Chemistry, Federal University of Mato Grosso do Sul,
Avenue Felinto Muller, 1555, PO Box 549, Campo Grande,
MS 79070-900, Brazil
e-mail: jorgeluizraposo@gmail.com

M. Z. Corazza
Faculty of Exact Sciences and Technology, Federal University
of Grande Dourados, Highway Dourados-Itahum, km 12,
PO Box 364, Dourados, MS 79804-970, Brazil

 
M. Trindade (ed.), *Increased Biodiesel Efficiency*, Green Energy and Technology,
https://doi.org/10.1007/978-3-319-73552-8_1

improvement of the environment, creation of new jobs, and a significant contribution to the domestic economy.

**Keywords** Biofuels · Biodiesel · Inedible oil

## 1.1 Introduction

Energy generated by the burning of fossil fuels represents about 80% of world production (Lee et al. 2017). Currently, Saudi Arabia, Russia, and the USA are the largest oil producers, with around 40% of world production (Panorama Internacional Petróleo 2017). Other countries in Central America, South America, Europe, and Africa also contribute a large share of oil production. However, due to the increase in consumption of this source by the energy sector, it is possible that the world market needs to evaluate alternative solutions to avoid the exhaustion of this resource. The importance of fossil fuels in energy production in the transportation sector is almost 98% (Baskar and Aiswarya 2016). According to the Energy Forum, world reserves of fossil fuels, coal, natural gas, and oil will be available for less than 10 decades (Pandit and Fulekar 2017), and a significant decline in production (30%) is expected by 2035 (Baskar and Aiswarya 2016).

In light of possible economic oscillations caused by the fall in oil production, the search for the development of alternative sources for energy production, whether through solar, wind, geothermal, tidal, or biomass energy, has been the subject of research by the scientific community (Nguyen et al. 2017). Among these possibilities, the production of energy using renewable sources obtained from the use of biomass is an economically viable alternative and is sufficiently promising, since the raw material for the production of biofuel can have local subsidies that contribute to the sustainable development and creation of new employment opportunities (Rajaeifar et al. 2017). The demands of the energy sector, not only in Brazil but also globally, have boosted this sector toward the implementation of programs to generate employment and income from the production of biodiesel to obtain environmental and social benefits (Silva et al. 2016a, b).

Among the renewable energy sources obtained from biomass, ethanol is undoubtedly the biofuel with the highest production level. However, the production of biodiesel has received great incentive as it has numerous advantages and may be a competitor and/or even substitute for diesel oil (Farobie and Matsumura 2017). Particulate emissions from diesel engines are a large environmental problem, since in June 2012 the World Health Organization classified diesel engine emissions as carcinogenic (Wang et al. 2017). Biodiesel, a biodegradable and low toxicity fuel, has gained special attention from a number of government agencies that have made it mandatory to add biodiesel to diesel (Batista et al. 2016).

Legislative acts of the European Union (EU) state the need for fossil fuels to be replaced by biofuels in the transport sector and that by 2020 at least 10% of the total energy consumption of the transport sector of each EU member should come from renewable energy sources (Budžaki et al. 2017).

## 1.2 Biofuels

Due to the constant drive toward reducing the use of fossil fuels due to increasing levels of undesirable gas emissions, the scientific interest in biofuels has received dedicated attention. In addition to creating new jobs, biofuels could improve the energy security of oil and gas imports (Azadi et al. 2017) and allow the inclusion of smallholders, rural development, and decentralization of energy systems (Hunsberger et al. 2017). Ambitious targets for renewable energy are being set by nations and international institutions in the Sustainable Development Agenda 2030 (German et al. 2017).

Bioenergy is considered a clean source of alternative energy that will play an important role as a substitute and complementary energy in the very near future (Su et al. 2017). The first important policy measures to promote the large-scale production and consumption of biofuels in the EU and the USA were legislated in 2003 and 2005, respectively (Azadi et al. 2017). In 2006, the production and consumption of biofuels increased rapidly worldwide (Hunsberger et al. 2017). Biofuel production from 2000 to 2009 was concentrated in some countries and sectors, such as sugarcane ethanol in Brazil and palm oil-based biodiesel in Colombia, Indonesia, and Malaysia (Hunsberger et al. 2017). In 2016, world biofuel production increased by 2% compared to 2015 and reached 135 billion liters (REN21 2017).

Gaseous biofuels are commonly used for the production of heat and energy, while liquids are used in the transportation industry. Biofuels include biomethanol (MeOH), bioethanol (EtOH), bio-dimethyl ether (DME), synthetic natural gas (bio-methane), Fischer–Tropsch (FT) fuels, and $H_2$ fuels (Sikarwar et al. 2017). They can be classified as first-, second-, third-, and fourth-generation biofuels (Sikarwar et al. 2017).

Biomethanol, also known as "wood alcohol," is among the ten most important chemicals produced in the global sphere (Swain et al. 2011). It can be produced by fermentation or by distillation of cultures containing sugar and starch, once it is converted to simple sugars and finally to biomethanol (Rodionova et al. 2017). This product is a liquid under normal conditions of temperature and pressure, which facilitates transport and handling. The main application of biomethanol as raw material is to produce formaldehyde (Swain et al. 2011).

Bioethanol is the most common biofuel in the world, particularly in Brazil and the USA. It is produced by fermentation of sugar obtained from wheat, corn, sugar cane, and molasses and can be used in blends with gasoline in any proportion or as a substitute (Ullah et al. 2015). This biofuel reduces emissions of lead, sulfur, CO, and particulate matter (Bahadar and Khan 2013). Ethanol used as fuel in Brazil reduces carbon emissions from 9.56 to 106 tons, contributing to a 15% reduction of total emissions in the country (Bahadar and Khan 2013).

Bio-dimethyl ether is a colorless, non-toxic, non-carcinogenic, and non-corrosive gas (Bhattacharya et al. 2013). It can be used as a diesel substitute because it has a high cetane number. Its combustion is smoke-free due to its low self-ignition temperature, high oxygen content, and absence of C–C bonds in the molecular structure (Bhattacharya et al. 2013). In addition to these intrinsic merits, bio-dimethyl ether has gained much attention as a fuel, since this biofuel can be produced from biomass (Delparish and Avci 2016).

Biogas, obtained from organic materials by anaerobic fermentation, has a composition of 40–70% methane, 60–30% carbon dioxide, and other gases (hydrogen sulfide, hydrogen, nitrogen, and carbon monoxide) (Bayrakci and Koçar 2012). Due to numerous applications in energy generation, such as heat, electricity, and vehicles fuel, the biogas industry has developed rapidly in recent years, especially in the EU and Asia (Xia et al. 2016).

Fischer–Tropsch synthesis is an important technology for the conversion of synthesis gas (CO and $H_2$ obtained from natural gas, coal, or biomass) in hydrocarbon products (Xiong et al. 2017). This route is effective in producing clean fuels and non-petroleum-based chemicals (Xiong et al. 2017). These fuels are of high quality due to the absence of aromatic compounds, sulfur, and nitrogen (Zhang et al. 2017).

Bio-hydrogen can be produced from various raw materials such as lignocellulosic biomass (Saba et al. 2015), whose only by-product is water and which does not cause damage to the environment (Chen et al. 2017). The production of hydrogen by biological methods can be divided into two categories: photo-energy-dependent fermentation and anaerobic fermentation (Singh and Wahid 2015). Bio-hydrogen can be used to generate energy, whether by combustion or by fuel cell systems. It is an important product that can also be used to produce fertilizers (e.g., urea) and other chemicals (Chen et al. 2017).

Biodiesel is one of the best sources of renewable energy, not only because it produces less impact on the environment but also because it offers the advantage of sustainability, being biodegradable and non-toxic (Hilioti et al. 2017). In the last decades, biodiesel is an alternative to fossil fuels and receives increasing attention (Rodionova et al. 2017). It can be produced from renewable biological materials and can replace petroleum diesel (Rodionova et al. 2017). It is produced by transesterification of various animal fats and vegetable oils, usually with the help of methanol or ethanol (Rodionova et al. 2017). It can be used in pure form (B100) or blends with petroleum diesel in any concentration in most diesel injection pumps (Bayrakci and Koçar 2012).

## 1.3 Raw Materials for Biodiesel Production

More than 300 oils have been investigated as raw material candidates for biodiesel production based on economic viability and availability (Mishra et al. 2017). The most frequently used raw materials for the production of biodiesel are soybean, coconut, palm, sunflower, canola (Rattanapoltee and Kaewkannetra 2014), cotton, corn, forage turnip, *tungue*, crambe (Chaves et al. 2010), bovine and ovine tallow, bird oil, oil of fish, almond, andiroba, babassu, barley, copra, cumaru, camelina, peanut, laurel, pequi, rice (Demirbas 2009), grape seeds (Ramos et al. 2009), insect fat, gross larvae fat (Nguyen et al. 2017), wheat, sorghum (Živković et al. 2017), and linseed (Bacenetti et al. 2017).

According to Živković et al. (2017), biodiesel can be classified into three categories depending on the type of raw material used in production: first, second, and third generation. First-generation biodiesel is produced from edible oils, second-generation biofuel is produced from inedible oils, and third-generation oils have been recently explored and include micro- and macro-species, such as algae (Baskar and Aiswarya 2016).

In addition to inedible species and frying residual oil, deodorized distilled soybean oil, which is a by-product in soybean oil refining, may be a potential low-cost feedstock for biodiesel production due to its high content of triglycerides and free fatty acids (Yin et al. 2017).

First-generation biofuels could have serious environmental impacts in relation to competition for farmland, water, emissions to the atmosphere, and loss of biodiversity. However, choosing a suitable harvesting type may contribute to the reduction of this impact (Bacenetti et al. 2017). The main barrier to the commercialization of biodiesel is the high added value, which makes it possible to use cheaper and non-edible oils in industrial-scale production (Yin et al. 2017). Table 1.1 presents different raw materials used in the production of biodiesel and their respective percentages of oil.

Due to doubtful competition in the use of traditionally edible oils used for food versus energy production purposes, the oils of the species *Cynara cardunculus* L., *Jatropha curcas*, *Brassica carinata*, and *Pongamia pinnata* (Martínez et al. 2014),

**Table 1.1** Oil content present in the main raw materials used to produce biodiesel (Baskar and Aiswarya 2016)

| Type of oil | Feedstock | Oil content % (w/w) |
|---|---|---|
| Edible | Soybean | 15–20 |
| | Rapeseed | 38–46 |
| | Sunflower | 25–35 |
| | Peanut oil | 45–55 |
| | Coconut | 63–65 |
| | Palm | 30–60 |
| Non-Edible | Jatropha seed | 35–40 |
| | *Pongamia pinnata* | 27–39 |
| | Neem oil | 20–30 |
| | Castor | 53 |
| Other sources | Rubber seed | 40–50 |
| | Sea mango | 54 |
| | Cotton seed | 18–25 |
| | Microalgae | 30–70 |

microalgae, residual cooking oil (Nguyen et al. 2017), microbial and insect oils (Živković et al. 2017), castor oil, tobacco seed oil, neem oil, rice bran oil, and rubber seed oil have been identified as promising raw materials for the production of oil and/or biodiesel (Negm et al. 2017).

Inedible crops can be planted in many parts of the world, including extensions of partially destroyed land. This activity can reduce the rate of deforestation and avoid competition with food crops (Agarwal et al. 2017). In addition, biodiesel derived from edible oils presents a higher cost than inedible oils, which generates a negative overall impact on the economy (Agarwal et al. 2017). Therefore, the scientific community has turned its interest to oils that are inedible due to the presence of some toxic components (Leung et al. 2010). Some inedible raw materials deserve to be highlighted and will be presented below.

### *1.3.1 Jatropha*

*Jatropha curcas* L. is a small tree that can be from 5 to 7 m in height. Native to South America, it belongs to the *Euphorbiaceae* family, which contains about 175 known species (Kumar et al. 2016). Jatropha seeds contain 27–59% oil (Rodrigues et al. 2016; Zhang et al. 2016), whose main fatty acids are linoleic, oleic, stearic, and palmitic, with a composition of 22.5% saturated fatty acids and 77.5% unsaturated fatty acids (Kumar et al. 2016).

The plant presents five roots, one deep primary and four peripheral ones, and exhibits articulated growth with morphological discontinuity; its branches contain latex (Laviola et al. 2017). Trees are deciduous, spreading their leaves in the dry season and flowering during the rainy season, usually twice a year during the summer and autumn (Laviola et al. 2017). Jatropha is a monoecious plant with inflorescences with separate female and male flowers, and the transport of the pollen by the wind is limited due to its weight and adhesiveness (Montes and Melchinger 2016). Therefore, insects play an important role in its pollination (Montes and Melchinger 2016).

It is a plant that adapts to poor soils and severe climatic conditions (Meloni et al. 2016) and can grow in annual rainfall conditions of up to 1500 mm; it is therefore capable of resisting long periods of drought (Naresh et al. 2012) if it develops in tropical and subtropical regions (Lim et al. 2016) and shows resistance to pests (Nalgundwar et al. 2016). Because it is an inedible species, there is no competition between the food and fuel production sectors (Meloni et al. 2016). This species contains a certain amount of fatty acids that are suitable for the production of biodiesel (Nalgundwar et al. 2016), besides being used in the production of soaps and bio-pesticides (Muldalkar et al. 2016).

### *1.3.2 Crambe*

Crambe (*Crambe abyssinica* Hochst) is an annual herb of the *Cruciferae* family and originated in the Mediterranean region (Ítavo et al. 2016). It has an oil content of approximately 38%, containing high levels of erucic acid (50–60%). Erucic acid is a monounsaturated fatty acid having 22 carbon atoms (Wazilewski et al. 2013) and is applied in the pharmaceutical, cosmetic, lubricant, and plastics industries (Santos et al. 2015; Lara-Fioreze et al. 2013). Crambe oil, in turn, is not edible, as it can cause damage to the heart and disturbance in the physiological processes of animals (No et al. 2013), besides being used as a corrosion inhibitor in the manufacture of synthetic rubber, nylon, and adhesives (Rogério et al. 2013). Recently, this oil has been applied in the production of biodiesel (Pitol et al. 2010).

About 35 species of this genus are known in four geographic regions: Macaronesia, Mediterranean, East Africa, and Eurosiberia–Southwest Asia (Tarikahya-Hacioğlu 2016). The plant may be 1.0–2.0 m in height, and the spherical seeds have diameters of 0.8–2.6 mm (Santos et al. 2015). Due to robustness, early cultivation, and resistance to drought, pests, and diseases, the maintenance of this crop can be done with the same implements as are used for traditional crops, such as soybean (Dos Santos et al. 2013). The ideal moisture content for the crambe crop should be 14% (Da Silva et al. 2012) and its life cycle is short, since flowering occurs between 40 and 50 days and seed maturity between 75 and 90 days (Chhikara et al. 2012).

Crambe is a winter crop with a low production cost and is easily adaptable to crop rotation (De Marins et al. 2018) with soybean, corn, wheat, and so on (Tavares et al. 2017). Its crop requires lower amounts of fertilizers and water compared to other oil crops, and the seed yield is between 1000 and 1500 kg/ha (Tavares et al. 2017). It is highly self-pollinated and presents only a low frequency within its own species (Carlsson et al. 2014).

### *1.3.3 Castor Bean*

The castor bean (*Ricinus communis*) is part of the *Euphorbiaceae* family and is believed to have originated in Africa, especially in Ethiopia, as there is a greater variety of wild and semi-cultivated castor bean in this region than in any other region of the world (Singh et al. 2015). Castor oilseeds (*Ricinus communis*) have an oil content of 50% and are used for the extraction of castor oil, which has many applications ranging from cosmetics to the biofuel industry (Hilioti et al. 2017). In addition, it is used in the manufacture of paints, varnishes, lubricants (Torrentes-Espinoza et al. 2017), synthetic nylon fibers, greases, waxes, and detergents (Omonov et al. 2017).

The main component of castor oil is ricinoleic acid (12-hydroxy-9-octadecenoic acid), which comprises between 86 and 92% of the fatty acids found in this oil

(Torrentes-Espinoza et al. 2017). The cultivation and processing of this species results in the production of significant amounts of dry residues with large amounts of ricinus, which is highly toxic to humans and animals, so it is necessary to perform a process before discarding them (Hilioti et al. 2017). A large proportion of castor oil is produced in countries with tropical climates such as India, China, and Brazil (Omonov et al. 2017).

Castor oil has been identified as an important inedible source for biodiesel production, and castor bean seeds require less fertilizer, pesticides, and water. This species can grow on marginal lands and provides an excellent opportunity for several regions to use their land more effectively (Akhabue et al. 2014). This species presents low costs of implementation and production and relative resistance to water stress, attributes that allow it to develop in adverse climatic and soil conditions (Bueno et al. 2017). Its productivity is 1180 kg of oil per hectare (Torrentes-Espinoza et al. 2017).

### *1.3.4 Pongamia Pinnata*

The *Pongamia pinnata* is a medium-sized tree that belongs to the family *Fabaceae* and subfamily *Papilionaceae* (Mariprasath and Kirubakaran 2016). Native to India and its subcontinents, it was also successfully introduced in the humid tropical regions of the world such as Australia, Newlands, China, and the USA (Mariprasath and Kirubakaran 2016). It exhibits rapid growth and can reach 20 m in height (Dwivedi et al. 2017). The oil content is 30–40%, and it is resistant to drought, heat, and salt conditions. This plant is used as a remedy for skin diseases, as pesticide, and as fertilizer, and the feasibility of its application in the biofuel industry has recently been recognized (Mariprasath and Kirubakaran 2016).

Small clusters of purple and pink flowers bloom in their twigs throughout the year, ripening in brown seed pods (Dwivedi et al. 2017). *Pongamia pinnata* contains a wide range of biologically active compounds such as flavonoids, terpenoids, phenols, saponins, alkaloids, and vitamins (Ambika and Sundrarajan 2015). Its seeds contain 46% oleic acid, 27% linoleic acid, 6% linolenic acid, and a small amount (0.1%) of low molecular weight fatty acids, such as lauric and capric acids (Malaikozhundan et al. 2017).

*Pongamia* has inedible oil and presents great potential for biodiesel production, being able to grow in several types of soils, and its productive life is more than 40 years (Dwivedi and Sharma 2015). Overall, many countries including Australia, India, and Hawaii have initiated economic modeling practices to encourage *Pongamia* planting programs as a source of renewable raw material for biodiesel (Radhakumari et al. 2017). It is used to control soil erosion by joining sand dunes due to its dense network of lateral roots (Sahu et al. 2017).

### 1.3.5 Waste Cooking Oil

Waste cooking oil is produced after the frying of edible oils (soybean oil, corn, sunflower, household fats, and so on), which alters the oils' physical and chemical properties. However, these changes do not affect biodiesel production (Xiang et al. 2017). In the biofuel production process, part of the WCO is transformed into biodiesel (80%) and glycerol (10%), while the 10% remaining is discarded (da César et al. 2017).

Besides the utility in the production of biodiesel, the residual oil can also be used in the manufacture of soaps and oil paints (da César et al. 2017). About 15 million tons of waste cooking oil (WCO) is discarded annually around the world (Dos Santos et al. 2017); it has a lower cost than refined oils and can also cause damage to the environment. Therefore, the production of biodiesel from WCO will reduce environmental pollution as well as the cost of production of this biofuel (Wang et al. 2017), as it is an undesirable waste from residences, restaurants, and industries (Manique et al. 2012) and also there is no competition with the food market (Atapour et al. 2014).

Initiatives for the use of WCOs are found in several countries, including Australia, China, Germany, Italy, Portugal, the UK, the USA, Austria, and Spain. Brazil also produces WCO biodiesel; however, its quantity in the energy matrix is still low (da César et al. 2017). In 2015, only 0.5% of biodiesel was produced from WCO and it is estimated that only 2.5% of the oil produced in the country is reused while the remainder is improperly discarded (da César et al. 2017). Biodiesel produced from jatropha oil and WCO is about four times cheaper than biodiesel produced from palm oil (Živković et al. 2017).

### 1.3.6 Microalgae

Microalgae are eukaryotic and prokaryotic photosynthetic microorganisms with a simple structure that allows them to grow rapidly. Under optimal growth conditions, some microalgae species can produce and accumulate 30–70% oil (Faried et al. 2017). Microalgae, bacteria, fungi, and yeasts are the most important microbial microorganisms, with microalgae being the best candidate for biodiesel production due to their higher photosynthetic efficiency, higher biomass production, and faster growth (Vicente et al. 2017).

Microalgae contribute to the reduction of $CO_2$ emissions, as they convert carbon into lipids, carbohydrates, and proteins. This oily source can be produced on land unsuitable for agriculture, such as deserts and coastal areas (Vicente et al. 2017), and therefore does not compete with food crops for arable land and water (Faried et al. 2017). Microalgae can produce hundreds of times more oil than soybeans, as shown in Table 1.2.

**Table 1.2** Comparison between the productions of oil by different raw materials (Faried et al. 2017)

| Crop | Oil yield (L/ha) |
|---|---|
| Canola | 1190 |
| Coconut | 2689 |
| Corn | 172 |
| Jatropha | 1892 |
| Microalgae[a] | 136.900 |
| Microalgae[b] | 58.700 |
| Oil palm | 5950 |

[a]70% oil (by wt) in biomass
[b]30% oil (by wt) in biomass

### 1.3.7 *Brassica Carinata*

*Brassica carinata*, commonly called Ethiopian mustard (Marillia et al. 2014), is a member of the *Cruciferous* family of plants and is composed of two genomes; one originating from the genome *Brassica nigra* B. and one from the genome *Brassica oleracea* C. (Canam et al. 2013). It was grown mainly in Ethiopia as a vegetable and oilseed crop (Li et al. 2009). It has valuable agronomic characteristics, such as better tolerance to heat, drought, disease resistance, and seed burning (Seepaul et al. 2016). The species *B. carinata* presents a strong competitiveness with weeds, and therefore, the use of herbicidal inputs (Marillia et al. 2014) can be minimized.

This species presents two types of seeds that are characterized by the color of their coats: brown and yellow. Generally, *B. carinata* with yellow seeds has larger seed size and higher oil and protein content than brown seed lines (Xin and Yu 2014). Typical *B. carinata* straw contains about 10% water, 16% hemicellulose, 29% cellulose, and 16% lignin (Stamigna et al. 2012).

The oil content represents about 40.0% (Seepaul et al. 2016); however, the seeds have a high glucosinolate content (>95%) that interferes with thyroid metabolism and damages vital organs (Marillia et al. 2014). Recently, this species was identified as one of the first seven crops destined for the production of biofuels, and tests indicated emissions without sulfur, with the reduction of 25% of the particles and 50% of the carbon (Marillia et al. 2014).

## 1.4 Biodiesel Production Processes

Biodiesel is defined as a monoalkyl ester of long-chain fatty acids derived from renewable lipid raw materials, produced by the transesterification of oils with short-chain alcohols, such as methanol or ethanol, in the presence of acid or basic catalysts (Lyra et al. 2010; Pereira et al. 2013) (Fig. 1.1). In the transesterification, a sequence of three consecutive reversible reactions occurs, in which the triglycerides are converted sequentially into diglycerides, monoglycerides, and glycerol.

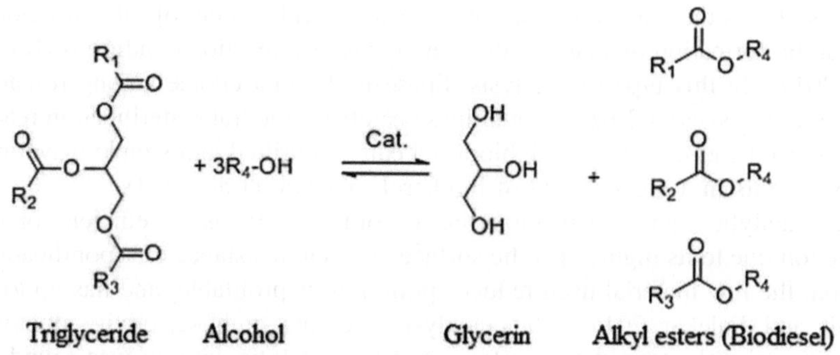


**Fig. 1.1** Typical transesterification reaction of triglycerides to monoalkyl esters conversion of fatty acids (biodiesel)

Each step produces an ester, where each molecule of triglyceride produces three molecules of ester (Yaakob et al. 2013).

In the transesterification reaction, methanol is the most widely used acyl acceptor because of its viability and economical accessibility compared to other alcohols (Wang et al. 2017). In the transesterification reaction, heterogeneous catalysts ($CaZrO_3$, $Al_2O_3$-SnO, Li/MgO, $Al_2O_3$/KI, KOH/$Al_2O_3$, KOH/NaY, and $K_2CO_3$ supported with alumina/silica) (Shu et al. 2010), homogeneous ($H_2SO_4$, NaOH, KOH), or enzymatic (lipase) (Baskar and Aiswarya 2016) are often used. In addition, the use of supercritical technology is observed in transesterification, whose reaction occurs in the absence of the catalyst (Baskar and Aiswarya 2016).

The current production of industrial biodiesel is performed using a homogeneous catalyst, and the main factor to be considered in the selection of this catalyst is the contents of water and free fatty acids present in the raw material (Živković et al. 2017). Although homogeneous catalysts are inexpensive, they need to be neutralized and removed from the reaction product and this process includes washing steps, which makes the process more complex (Živković et al. 2017). The basic catalysts are the most common, since the process is faster and the reaction conditions are moderate (Vicente et al. 2004). The acid-catalyzed reaction requires large amounts of alcohol to provide a good yield of esters (Baskar and Aiswarya 2016).

Heterogeneous acid catalysis offers some advantages over the homogeneous type, such as easy separation of the reaction mixture and reuse (Rajaeifar et al. 2017). Like homogeneous catalysts, heterogeneous catalysts are classified into two types: acid and basic (Yaakob et al. 2013). The heterogeneous catalysts are not hydroscopic solids and can be easily removed by filtration; they catalyze the transesterification of oils and the esterification of contaminated acids (Negm et al. 2017). Highly effective heterogeneous catalysts derived from residues such as eggshell, oyster shell, shrimp shell, crab shell, and mollusk waste shells were used in biodiesel production (Chakraborty et al. 2011).

Enzymatic production is possible using intracellular and extracellular lipases (Ranganathan et al. 2008). In comparison with chemical catalysis, enzymatic catalysis allows the conversion of high free fatty acid content of oils into biodiesel without the formation of soaps or the need for severe reaction conditions (Rajaeifar et al. 2017). In this type of catalysis, lipase is the first choice among researchers (Baskar and Aiswarya 2016). When lipase catalyzes the transesterification reaction, glycerol can be easily recovered, biodiesel can be purified in a simple way, and the process results in a higher yield of biodiesel (Yaakob et al. 2013).

Nanocatalytic transesterification has become the focus of efficient biodiesel production due to its highly specific surface area and resistance to saponification. In addition, the raw material used reduces pollution, is profitable, and has no toxicity (Pandit and Fulekar 2017). Nanocatalysis presents high selectivity due to the nano-dimensional pores at the surface, and the catalytic activity was found to be unchanged even after the second cycle (Baskar and Aiswarya 2016).

Supercritical transesterification can be performed without using any catalyst, allows easier separation of the products, and has faster reactions (Campanelli et al. 2010). In addition, the water and free fatty acid contents do not affect the yield of esters (Campanelli et al. 2010). The supercritical conditions require high temperature and pressure and are some of the biggest obstacles to commercialization (Tan et al. 2010). In addition, this process uses a large excess of alcohol and has a high energy consumption and a high capital cost (Yaakob et al. 2013). The transesterification of triglycerides and the esterification of fatty acids occur simultaneously (Farobie and Matsumura 2017).

The final yield of the transesterification reaction is influenced by several important parameters, such as the reaction temperature, type of catalyst, type and structure of the alcohol, amount and concentration of the catalyst, reaction time, and technical aspects related to the system (Martínez et al. 2014). The glycerol obtained in the transesterification reaction after the purification step can be used for various commercial purposes in the production of cosmetics, pharmaceuticals, and food or can be recycled as a source of catalyst (Agarwal et al. 2017). The treatment step of the methyl ester includes evaporation of the alcohol, neutralization, purification, and distillation (Živković et al. 2017). The remainder of the catalyst, soap, salts, alcohol, and free glycerol are removed from the biodiesel by washing with water (Agarwal et al. 2017).

In addition to transesterification, techniques such as microwave-assisted, ultrasonic-assisted, in situ transesterification, and membrane-assisted techniques (Rajaeifar et al. 2017) as well as pyrolysis and micro-emulsification have been used in the production of biodiesel (Budžaki et al. 2017).

The production of biodiesel includes the step of lipid isolation of the raw materials with organic solvents in order to avoid interference in the synthesis of methyl esters of fatty acids (Lee et al. 2017). Another process used as an alternative form of this isolation is supercritical fluid extraction (Lee et al. 2017).

## 1.5 Biodiesel Properties and Environmental Impacts

Biodiesel has four to eight methyl esters as the main constituents, such as myristate (C15:0), palmitate (C17:0), stearate (C19:0), oleate (C19:1), linoleate (C19:2), and linolenate (C19:3) (Thangaraja et al. 2016). The variations in the concentrations of saturated and unsaturated methyl esters in biodiesel due to the different raw materials used influence their physicochemical properties and consequently the characteristics of the engines (Mishra et al. 2017). Biodiesel containing a high content of saturated molecules presents a higher cetane number and energy content than biodiesel containing unsaturated molecules (Tongroon et al. 2017). Similarly, the cloud point and the cold point connection are larger for a bigger fraction of saturated fatty acids (Tongroon et al. 2017).

This biofuel is a clear liquid with a color ranging from light yellow to dark yellow, boiling point greater than 200 °C, flash point between 145 and 175 °C, distillation range of 195 and 325 °C, and vapor pressure (mm Hg at 22 °C) of less than 5.0 (Yaakob et al. 2013).

Biodiesel has a higher flash point, lubricity, cetane number, biodegradability, non-toxicity (Gonca and Dobrucali 2016), and renewability (Zhang et al. 2012). The degradation of biodiesel can reach about 90% after 30 days, while that of diesel reaches only 25% under the same conditions (Živković et al. 2017). The combustion emissions of biodiesel are free from heavy metals such as Cd, Co, Cu, Pb, V, and Ni and other pollutants such as $NO_x$ and $SO_x$ (Orozco et al. 2014). In addition, its use eliminates the emission of black smoke and reduces the emission of particulate material (Nogueira and Do Lago 2011).

The benefits of biodiesel still include compatibility with existing engine cycles, increased storage and transport safety, and the easy formation of blends with diesel oil (Dhamodaran et al. 2017). The production technology is easy to use and economically viable, and the existing infrastructure can be used (Agarwal et al. 2017).

With the use of biodiesel, there is a 78% reduction in $CO_2$ emissions compared to diesel oil, since plant species use atmospheric $CO_2$ to perform photosynthesis (Dhamodaran et al. 2017). Biodiesel has an oxygen mass of 10 to 12% in the molecular structure, which improves the combustion efficiency and reduces emissions of carbon monoxide (CO) and hydrocarbons (Gülüm and Bilgin 2015) (Table 1.3). It is a non-flammable and non-explosive biofuel (Xue et al. 2016). It is beneficial for reservoirs, lakes, marine life, and other environmentally sensitive places (Gülüm and Bilgin 2015).

**Table 1.3** Comparison of greenhouse gas emissions by diesel fuels, soybean biodiesel, and microalgae (Živković et al. 2017)

| Fuel | Type of GHG (g/MJ) | | |
|---|---|---|---|
| | $CO_2$ | $CH_4$ | $N_2O$ |
| Conventional diesel | 14.69 | 2.48 | 0.07 |
| Soybean biodiesel | −72.73 | 0.42 | 0.58 |
| Microalgae | −59.49 | 0.74 | −16.54 |

In addition to the above-mentioned advantages, biodiesel was used to remove oil from affected littoral or sediments and to stimulate oil degradation when applied to contaminated sand in laboratory tests (Fernández-Álvarez et al. 2007). Its use as an alternative solvent in polymerization techniques has been reported (da Costa et al. 2016), and it can also be used as a cleaning and preserving solvent for processed metal (Shah et al. 2016). Two significant advantages in the biodiesel life cycle are the reductions of greenhouse gas emissions and of the national dependence on fossil energy (Živković et al. 2017).

Biodiesel has some disadvantages, such as high viscosity, low calorific value and volatility, low oxidation stability, and hydroscopicity, and if used as a solvent it can cause corrosion of components, attacking some plastic materials used for hoses and coatings (Gonca and Dobrucali 2016). In addition, the production of biodiesel from vegetable oils and animal fat presents a higher cost compared to diesel; up to 50% higher (Gülüm and Bilgin 2015). This high cost for the production of biodiesel is the biggest obstacle to commercialization (Gülüm and Bilgin 2015).

The use of biodiesel based on methyl esters has a negligible effect on the unregulated emission of acrolein, propanol, and acetone compared to diesel (Agarwal et al. 2017). However, the use of biodiesel based on ethyl esters increases the emission of unregulated species such as acetaldehyde, acrolein, propanol, and acetone and, to a lesser extent, formaldehyde compared to diesel (Agarwal et al. 2017).

## 1.6 Blends of Diesel Fuel and Biodiesel

The biodiesel can be added to diesel fuel, forming blends for application in diesel engines. The composition of a biodiesel/diesel blend is indicated by "Bx," where "x" represents the percentage of biodiesel (v/v) present (Corach et al. 2016). As long as these products are new, they are completely miscible and the composition of the blends can be chosen according to the requirements of the engine (Corach et al. 2016).

Due to environmental and supply concerns, regulations in many countries state that diesel should be mixed with biodiesel of a certain value (Corach et al. 2016). In addition, in many developing countries, the small-scale production of biodiesel using local feedstock has become an economically viable alternative to the operation of rural machines using biodiesel blends and diesel (Corach et al. 2016). In the USA and Canada, mandatory addition of 20% biodiesel to diesel is carried out; in the Czech and Slovak republics, the required addition is 30%, and in the EU and Argentina, it is 5% (Roveda et al. 2016). Biodiesel can be used in blends with diesel because it has similar properties (Table 1.4).

According to the US Environmental Protection Agency (EPA), the use of B20 [20% (v/v) biodiesel and 80% (v/v) diesel] can reduce particulate emissions by 10.1% compared to the use of 100% diesel oil and the emission of carbon monoxide and hydrocarbons can be reduced by 21.1 and 11.0%, respectively (Liu et al. 2017).

**Table 1.4** Comparison physicochemical parameters for diesel and biodiesel based on American Society for Testing and Materials (ASTM) (Helwani et al. 2009)

| Property | Diesel | Biodiesel |
|---|---|---|
| Standard number | ASTM D975 | ASTM D6751 |
| Composition | Hydrocarbon ($C_{10}$–$C_{21}$) | Fatty acid methyl Ester ($C_{12}$–$C_{22}$) |
| Specific gravity (g/mL) | 0.85 | 0.88 |
| Flash point (K) | 333–353 | 373–443 |
| Cloud point (K) | 258–278 | 270–285 |
| Pour point (K) | 243–258 | 258–289 |
| Water (vol.%) | 0.05 | 0.05 |
| Carbon (wt%) | 87 | 77 |
| Hydrogen (wt%) | 13 | 12 |
| Oxygen (wt%) | 0 | 11 |
| Sulfur (wt%) | 0.05 | 0.05 |
| Cetane number | 40–55 | 48–60 |

According to Arad et al. (2017), most researchers agree that there is a 50% reduction in emissions of particulate matter from diesel engines when biodiesel is used. Several factors have been suggested to explain the reduction of emissions of particulate matter by the use of biodiesel or blends of biodiesel and diesel: the oxygen content in biodiesel molecules, the absence of aromatic compounds in biodiesel, the different oxidative reactivity between the particles of diesel and biodiesel, the absence of sulfur in biodiesel, and the long start time of biodiesel combustion (Arad et al. 2017).

## 1.7 Biodiesel Quality Parameters

In Brazil, the criteria for quality control of biodiesel are established by the National Petroleum, Natural Gas and Biofuels Agency (ANP). The ANP, in accordance with Resolution no. 45 of August 25, 2014, defined biodiesel as a fuel composed of alkyl esters of carboxylic acids of long chain, produced from the transesterification and/or esterification of greases, of fats of vegetable origin or animal, and that meets the specification contained in the Technical Regulation, an integral part of this Resolution (ANP 2014). The standard procedures established by the ANP have the purpose of verifying the efficiency of the biofuel purification stage (Ferreira et al. 2015).

Similarly, the EU and the USA established standards for biodiesel through Resolutions EN 14214 and ASTM D6751, respectively (Batista et al. 2016) (Table 1.5). It is necessary to improve the quality of biodiesel to avoid engine damage and environmental impacts, as there is a great interest in this biofuel and dedicated efforts to achieve a high technological level for its production

**Table 1.5** Physicochemical parameters used to biodiesel characterization in accordance with current legislation (Živković et al. 2017)

| Property, unit | EN 14214 | | ASTM 6751 | |
|---|---|---|---|---|
| | Limit | Test method | Limit | Test method |
| Density at 15 °C, $g/cm^3$ | 0.86–0.90 | EN ISO 3675<br>EN ISO 12185 | – | – |
| Viscosity at 40 °C, $mm^2/s$ | 3.50–5.00 | EN ISO 3104 | 1.9-6.0 | D445 |
| Flash point, °C | 120 min | EN ISO 3679 | 130 min | D93 |
| Cetane number | 51 min | EN ISO 5165 | 47 min | D613 |
| Sulfur content, mg/kg | 10.0 max | EN ISO 20846<br>EN ISO 20884 | 15.0 max | D5453 |
| Phosphorus content, mg/kg | 10.0 max | EN 14107 | 10.0 max | D4951 |
| Metals I (Na and K), mg/kg | 5.0 max | EN 14538 | – | – |
| Metals II (Ca and Mg), mg/kg | 5.0 max | EN 14538 | – | – |
| Water content, mg/kg | 500 max | EN ISO 12937 | 500 max | D2709 |
| Acid number, mg KOH/g | 0.50 max | EN 14104 | 0.80 max | D664 |
| Free glycerol, % mass | 0.02 max | EN 14105; EN 14106 | 0.02 max | D6584 |
| Total glycerol, % mass | 0.25 max | EN 14105 | 0.24 max | D6584 |
| Sulfated ash content, % mass | 0.02 max | ISO 3987 | 0.02 max | D874 |
| Methanol content, % mass | 0.20 max | EN 141110 | – | – |
| Monoglycerides, % mass | 0.80 max | EN 14105 | – | – |
| Diglycerides, % mass | 0.20 max | EN 14105 | – | – |
| Triglycerides, % mass | 0.20 max | EN 14105 | – | – |
| Ester content, % mass | 96.5 min | EN 14103 | – | – |
| Linolenic acid methyl ester, % mass | 12.0 max | EN 14103 | – | – |
| Carbon residue, % mass | – | – | 0.05 max | D4530 |
| Iodine value, g $I_2$/100 g | 120 max | EN 14111 | – | – |
| Oxidation stability at 110 °C, time | 6 min | EN 14112 | – | – |
| Copper corrosion (3 h at 150 °C), degree of corrosion | No. 1 | EN ISO 2160 | No. 3 | D130 |
| Distillation 90% recovered, °C | – | – | 360 max | D1160 |

(Amais et al. 2012). The quality control of biodiesel is an important issue for its commercialization (Batista et al. 2016).

The determination of inorganic constituents is also important in the evaluation of biodiesel quality control, because in high concentrations, the metals can cause environmental damage and problems in engines (Korn et al. 2010), accelerate the decomposition of the biofuel, catalyze the oxidation reaction, reduce the thermal stability and efficiency, contribute to corrosion of internal parts of the engines, and compromise the catalytic system (Ferreira et al. 2015).

Some metals are introduced during the production process (Na, K, Ca, and Mg); others are present in the raw material (P, S, and Zn) or are used as additives (Si, Mn, Cr, Fe, and Ni) (Almeida et al. 2014) to increase the performance of biodiesel (Quadros et al. 2011). In addition, they can be introduced into biodiesel during oil extraction, washing, refining, transport, and storage (Ferreira et al. 2015). The metals Fe, Ni, Mn, Co, and Cu are commonly found in metallurgical barrels and tanks that store biodiesel (Almeida et al. 2014). In addition to metals, residual alcohol, soaps, free glycerol, and residual glycerides are contaminants that also impair the performance of biodiesel in diesel engines (Squissato et al. 2017).

## 1.8 Biodiesel in Brazil

During the 1940s, Brazil carried out the first attempt to exploit energy from oils and fats in internal combustion engines. There are reports of studies on the use of vegetable oils derived from babassu, coconut, chestnut, and cotton (Pousa et al. 2007). In 1975, the National Alcohol Program (PROALCOOL) was created to reduce imports of petroleum with the production of ethanol from sugarcane. This program presented positive environmental, social, and economic results (Goldemberg et al. 2004).

In 1980, the PRO-ÓLEO program, whose purpose was to incorporate up to 30% vegetable oils, was created during the oil crisis. However, later, with the drop in the price of oil on the international market, this program was abandoned in 1986 (Pousa et al. 2007). The first incentive for biodiesel production came in 2002 with the Brazilian Program of Technological Development for Biodiesel (PROBIODIESEL), which was responsible for the official introduction of biodiesel into the Brazilian energy matrix (De Oliveira and Coelho 2017).

In 2004, Brazil developed the National Biodiesel Production and Usage Program (PNPB), in which it determined the mandatory use of this biofuel in blends with diesel with a progressive percentage (Oliveira et al. 2017). The principle of the program was based on social inclusion and regional development, through political incentives for the production of biodiesel from oilseeds derived from family farming (Rico and Sauer 2015).

In 2005, a PNPB instrument entitled Social Fuel Stamp based on the social and productive inclusion of family agriculture, regional development, and sustainable production of renewable raw materials was introduced (Selo Combustível Social 2015).

Biodiesel production in Brazil increased from 736 $m^3$ in 2005 to approximately 2.7 million $m^3$ in 2011 (Castanheira et al. 2014). From January 2008 to November 2014, the mandatory amount of biodiesel in blends with diesel increased from 2% (B2) to 7% (B7) (Oliveira et al. 2017), and an increase in biodiesel consumption in Brazil was quickly seen (Quadros et al. 2011).

Currently, Brazil is one of the world leaders in the production of bioethanol and biodiesel (Silva et al. 2016a, b). Different oilseeds used in the production of

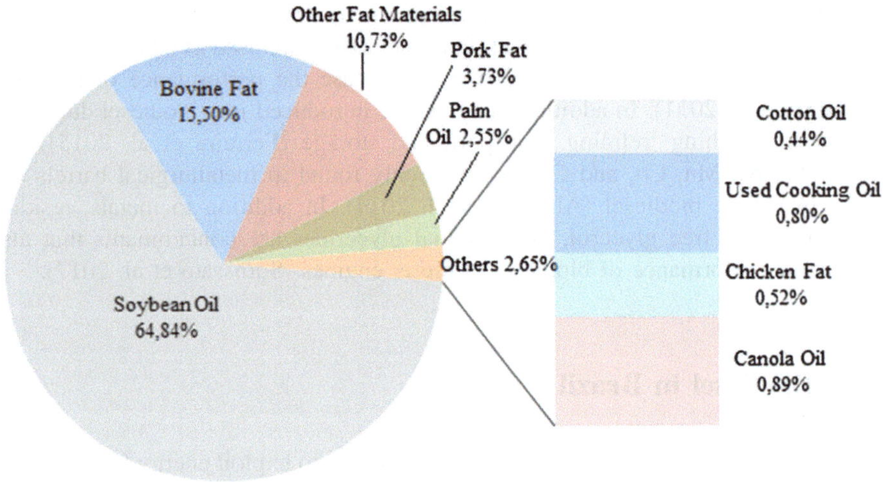


**Fig. 1.2** Raw materials used to produce biodiesel in Brazil in February 2017 (Bulletin 2017)

biodiesel can be grown in Brazil, due to the great extension of land suitable for agriculture and the great variety of climate and soil (Quadros et al. 2011). Figure 1.2 shows the mainly raw materials used in 2017.

The biodiesel currently produced in Brazil is consumed in the country. Until 2012, soybean biodiesel, produced mainly in the south and midwest, accounted for more than 85% of the cumulative production of biodiesel under the PNPB program (Naylor and Higgins 2017). The main policy objectives of promoting renewable energy and agricultural interests combined with the efficient use of by-products from the soy sector allowed Brazil to emerge as the third largest world producer of biodiesel after the US and Germany (Naylor and Higgins 2017).

As mentioned earlier, different raw materials can be used for the production of biodiesel. However, those most frequently exploited in Brazil are presented in Table 1.6.

The Decennial Energy Extension Plan 2023 foresees that in the year 2023, there will be a domestic demand for six billion liters of B7 and 8.6 billion liters of the B10 blend (De Oliveira and Coelho 2017). The incentive to produce biodiesel, regardless of its source of raw material, aims to reduce diesel pollution more and more, as declared in the National Energy Plan 2030, created in 2007 by the Energy Research Company (EPE) (De Oliveira and Coelho 2017).

Biodiesel is already a reality in the country and guarantees Brazil a leading position in relation to the rest of the world. Together, ethanol and biodiesel strengthen the participation of biofuels in the national energy matrix and the image of Brazil as a country that values the diversity of energy sources (Biodiesel 2017). In 2017, it is mandatory to add 8.0% biodiesel to the diesel fuel sold to the final consumer (Shimamoto and Tubino 2016), and in March 2019, that content will be raised to 10% (Brasil 2016).

**Table 1.6** Main raw materials used for biodiesel production in Brazil by region in February 2017 (Boletim 2017)

| Feedstock | North | Northeast | Midwest | Southeast | South |
|---|---|---|---|---|---|
| | % | | | | |
| Soybean oil | – | 47.02 | 75.65 | 26.72 | 65.35 |
| Bovine fat | 100 | 20.61 | 4.07 | 52.34 | 18.59 |
| Cotton oil | – | 5.77 | – | – | – |
| Other fat | – | 7.08 | 20.05 | 1.37 | 2.82 |
| Used cooking oil | – | 0.03 | 0.12 | 7.26 | 0.14 |
| Pork fat | – | – | 0.09 | 0.01 | 9.72 |
| Chicken fat | – | – | 0.02 | 1.18 | 1.05 |
| Palm oil | – | 19.49 | – | 11.11 | – |
| Rapeseed/canola oil | – | – | – | – | 2.34 |

## 1.9 Production of Biodiesel in Other Countries

Several countries have invested in the production and commercial viability of biodiesel through production units with different capacities, mainly in Europe (France, Austria, Germany, Belgium, the UK, Italy, the Netherlands, Finland, and Sweden), North America (USA), and Asia (Japan) (Lobo et al. 2009).

Many countries have introduced biodiesel into their energy matrices, and it has been used in compression ignition engines designed for diesel oil with few or no modifications (Silva et al. 2016a, b). On a global scale, biodiesel is expected to account for 70% of the growth in demand for transport fuel by 2040 (Naylor and Higgins 2017). The volume of biodiesel produced from 2002 to 2013 grew by a factor of 15 (De Oliveira and Coelho 2017) (Fig. 1.3). The use of different raw materials for biodiesel production in each country depends on local availability and is presented in Table 1.7.

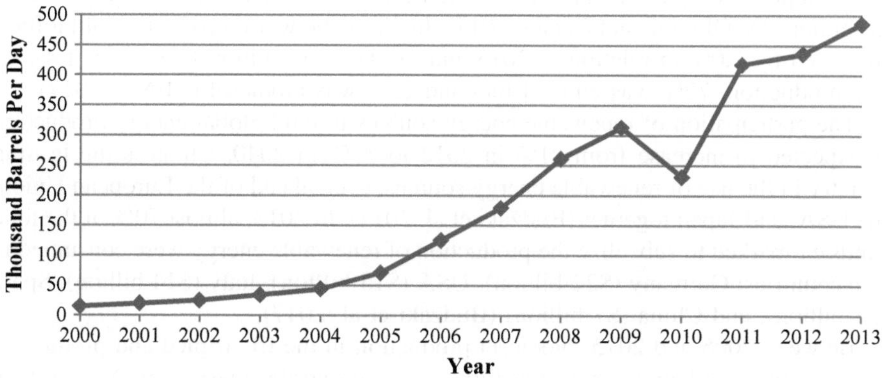


**Fig. 1.3** World production of biodiesel from 2000 to 2013 (De Oliveira and Coelho 2017)

**Table 1.7** Main raw materials used by biodiesel producing countries (Agarwal et al. 2017)

| Country | Feedstock |
|---|---|
| Australia | Waste cooking oil, animal fat |
| Brazil | Soybean, palm, castor, cottonseed |
| Canada | Canola, animal fat |
| China | Jatropha, waste cooking oil |
| Finland | Rapeseed, animal fat |
| France | Rapeseed, sunflower |
| Germany | Rapeseed |
| India | Jatropha, karanja |
| Indonesia | Palm, jatropha |
| Italy | Rapeseed |
| Japan | Waste cooking oil |
| Korea | Waste cooking oil |
| Malaysia | Palm |
| Mexico | Animal fat, waste cooking oil |
| New Zealand | Waste cooking oil, animal fat |
| Philippines | Coconut, jatropha |
| Russia | Rapeseed, soybean, sunflower |
| Spain | Sunflower |
| Sweden | Rapeseed |
| Thailand | Palm, coconut, jatropha |
| UK | Rapeseed, waste cooking oil |
| USA | Soybean, waste cooking oil |

Since 2000, the EU has expanded biodiesel consumption, and the USA, Brazil, Indonesia, and Argentina are also big producers of biodiesel. Other countries, such as Thailand, Colombia, Malaysia, and Canada, have significant market shares (Naylor and Higgins 2017), as shown in Table 1.8.

In 2013, world production and consumption of biofuels in transport increased by 7.0%, representing more than 116 billion liters, of which biodiesel represented 26 billion (De Oliveira and Coelho 2017). In 2016, the world production of biofuels increased by 2.0% in relation to 2015 and reached 135 billion liters (Fig. 1.4). Of this production, 72% was ethanol fuel and 23% was biodiesel (REN21 2017).

The participation of renewable energy sources in total global energy production is expected to increase from 21% in 2012 to 33% in 2040. China is the leading country in the use of renewable energy sources, even ahead of the European Union, the USA, and Japan together (Budžaki et al. 2017). In 2013, almost 70% of the total funds earmarked to subsidize the production of renewable energy were consumed in five countries: Germany ($22 billion), USA ($15 billion), Italy ($14 billion), Spain ($8 billion), and China ($7 billion) (Budžaki et al. 2017).

Between 2005 and 2015, biodiesel production in the EU tripled and production capacity increased more than five times, as large agribusiness firms have set up biodiesel refineries for imports of potentially competitive raw materials (Naylor and

**Table 1.8** Biodiesel production in different countries in 2015 (Naylor and Higgins 2017)

| Country | 2015 Biodiesel production, billion liters | World total (%) |
|---|---|---|
| Argentina | 2.1 | 7.0 |
| Brazil | 4.0 | 13 |
| Canada | 0.3 | 1.0 |
| Colombia | 0.6 | 2.0 |
| EU | 13.5 | 43 |
| Indonesia | 1.2 | 4.0 |
| Malaysia | 0.6 | 2.0 |
| Thailand | 1.2 | 4.0 |
| USA | 4.8 | 15 |
| Rest of World | 3.3 | 10 |
| World | 31.6 | 100 |

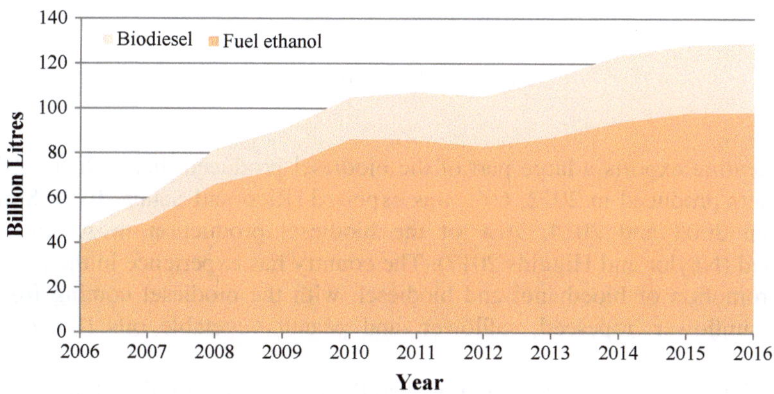


**Fig. 1.4** World production of biofuels biodiesel, etanol, and hydrotreated vegetable oil (HVO) from 2006 to 2016 (REN21 2017)

Higgins 2017). In 2016, the EU market accounted for 8.0 billion liters (Table 1.9) of biodiesel production and its share of the global market was around 26% (REN21 2017). The EU is encouraging the research and implementation of second- and third-generation fuels (Agarwal et al. 2017).

Biodiesel production in the USA accounted for 18% of world production in 2016, with a production of 5.5 billion liters (REN21 2017). The US Renewable Fuel Standard (RFS), introduced in 2005 and strengthened in 2007, requires the use of at least 136 billion liters of renewable fuels in the USA in the transport sector by 2022 (Naylor and Higgins 2017). Through the RFS and tax incentives, the price of biodiesel could be lower than that of diesel; however, currently the price of biodiesel is higher (Rico and Sauer 2015). The USA is the world's largest producer of soybean biodiesel, followed by Brazil (Rico and Sauer 2015).

**Table 1.9** Top 14 biodiesel producing countries and EU-28 in 2016

| Country | Biodiesel (billion liters) |
|---|---|
| Argentina | 3.0 |
| Belgium | 0.5 |
| Brazil | 3.8 |
| Canada | 0.4 |
| China | 0.3 |
| Colombia | 0.5 |
| EU-28 | 8.0 |
| France | 1.5 |
| Germany | 3.0 |
| Indonesia | 3.0 |
| Poland | 0.9 |
| Spain | 1.1 |
| Thailand | 1.4 |
| United Kingdom | 0.2 |
| United States | 5.5 |
| World total | 30.8 |

Adapted (REN21 2017)

Argentina exports a large part of the biodiesel produced there. Of the 2.78 billion liters produced in 2012, 60% was exported (Rico and Sauer 2015). Similarly, between 2008 and 2014, 70% of the biodiesel production in Argentina was exported (Naylor and Higgins 2017). The country has experience in the production and promotion of bioethanol and biodiesel, with the biodiesel coming from soybean, sunflower, rapeseed, safflower, and peanut vegetable oils (Lamers et al. 2008).

Thailand developed the Alternative and Renewable Energy Development Plan in 2012 and the government established a B7 mix mandate, with pilot projects for trucks and fishing boats with blends ranging from B10 to B20. It is the third largest producer of palm oil after Indonesia and Malaysia and has developed its biodiesel sector almost exclusively from that oil (Naylor and Higgins 2017).

The biodiesel industry in Indonesia has grown rapidly. The Indonesian Ministry of Energy and Mineral Resources (MEMR) plan seeks to diversify domestic biodiesel consumption beyond the transportation sector to include the industrial and electrical sectors. Fuels are subsidized and production costs are the lowest compared to other countries (Rico and Sauer 2015). According to the development of biofuels, the Indonesian government established a mandate to replace 15 and 20% of fossil fuels with bioethanol and biodiesel, respectively, by 2025 (Indrawan et al. 2017). The use of palm biodiesel in Indonesia generates independence from fossil fuels (Indrawan et al. 2017).

After Indonesia, Malaysia is the second largest producer of palm oil, which is intended to domestic demand for oil and fats as well as export (Rico and Sauer 2015). Indonesia and Malaysia have plans to emerge as the main oil producers

(20 and 14%, respectively) by 2021. It is expected that in 10 years the total production of palm oil of the two countries will increase by 37%. As it provides independence in the supply of raw materials such as palm oil, the country has sufficient reasons to support the production and use of biodiesel on an industrial scale. A biodiesel blend (B5) was officially introduced by the National Biofuel Policy at the end of 2009 (Ashnani et al. 2014). Currently, the country meets its B7 biodiesel mandate (Naylor and Higgins 2017).

The First National Energy Conference was convened in 1998 to promote the use of new and clean energy in Taiwan and the development of biodiesel was one of the projects planned at that meeting. In 2005, the second meeting was held and it was decided to increase the contribution of biodiesel to 100 million liters per year by 2010 and 150 million liters by 2020 (Huang and Wu 2008).

The Chinese biodiesel industry is in its primary stage with a global trend of biodiesel development. China has a brilliant potential in biodiesel production capacity. Currently, three types of raw materials are used for the production of biodiesel: rapeseed oil, cottonseed oil, and jatropha oil, and these raw materials will change to residual frying oil and subsequently microalgae in the future (Xu et al. 2006).

Biodiesel is emerging in Africa with large-scale investment projects for the beginning of its commercial production in several countries, including Ghana, Zambia, Liberia, Tanzania, Ethiopia, Nigeria, Senegal, Kenya, and South Africa. Currently the main inedible resource for biodiesel production in Africa is jatropha, but studies indicate that inedible oils such as jojoba, mahua, castor, and karanja present adequate properties for biodiesel and are widespread in many parts of the continent (Yang et al. 2014).

In Canada, the promotion of renewable fuels is an important component in reducing greenhouse gas emissions and mandates on fuel blends, and federal production subsidies have been used to stimulate growth in biofuel production. Canola accounts for more than half of the raw materials used in domestic production of biodiesel with recycled vegetable oil, and animal fat has a secondary paper (Naylor and Higgins 2017).

The first studies related to biodiesel in Turkey began in 1931 at the Agriculture Congress. Turkey has great potential for cotton and sunflower production and produced 17,729 tons of biodiesel in 2012 and 21,876 tons in 2013. According to the legislation, it became mandatory that diesel provided in the trade as fuel oil should contain 1.0% ester methyl fatty acid from January 1, 2014; 2.0% from January 1, 2015; and 3.0% from January 1, 2016 (Eryilmaz et al. 2016).

In the production of biodiesel in Peru, two types of potential raw materials were identified, namely palm oil and jatropha. In 2009, blends of 2.0% biodiesel with diesel started to gradually increase to 5.0% biodiesel by 2011. The development of the biodiesel industry in Peru is emerging, and the Romero group has vertically integrated biodiesel processing for its palm oil production facilities in the Amazon (Quintero et al. 2012).

Primary energy demand in Iran is expected to rise to an annual average of 2.6% from 2003 to 2030, 5.0% lower than in the past decade. Canola, cotton, and soybeans are the most favorable biodiesel production sources in Iran; however, sesame, olive, sunflower, safflower, almond, corn, and coconut are other potential oils. Total oilseed production in Iran from 2007 to 2008 was 3,678,540.43 tons (Ardebili et al. 2011).

Colombia is the fourth largest producer of palm oil in the world, and the country has the potential to boost biodiesel production to the local market. To promote the production of biodiesel, the Colombian government has promoted laws and policies establishing tax exemptions and subsidies as well as mandates for blends of diesel and biodiesel (Espinoza et al. 2017). Colombia's national blending mandate is set at B10 (and B8 for some remote rural areas) and the industry operated at around B7.9 in 2015 (Naylor and Higgins 2017).

Based on extensive research conducted at agricultural research centers, it was decided to use jatropha as the main raw material for the biodiesel program in India. However, commercial production of biodiesel in that country has not started (Khan et al. 2009).

## References


Agarwal AK, Gupta JG, Dhar A (2017) Potential and challenges for large-scale application of biodiesel in automotive sector. Prog Energ Combust 61:113–149

Akhabue CE, Iworah JC, Aisien FA (2014) Effect of selected metal contaminants on the stability of castor oil methyl ester. J Fuels 2014:1–6

Almeida JMS, Dornellas RM, Yotsumoto-Neto S, Ghisi M, Furtado JGC, Marques EP, Aucélio RQ, Marques ALB (2014) A simple electroanalytical procedure for the determination of calcium in biodiesel. Fuel 115:658–665

Amais RS, Garcia EE, Monteiro MR, Nóbrega JÁ (2012) Determination of Ca, Mg and Zn in biodiesel microemulsions by FAAS using discrete nebulization. Fuel 93:167–171

Ambika S, Sundrarajan M (2015) Plant-extract mediated synthesis of ZnO nanoparticles using *Pongamia pinnata* and their activity against pathogenic bacteria. Adv Powder Technol 26:1294–1299

ANP (2014) Agência Nacional de Petróleo, Gás Natural e Biocombustíveis. Resolução nº 45, de 25 de agosto de 2014

Arad A, Sher E, Enden G (2017) Modeling soot formation in diesel-biodiesel flames. Fuel 206:437–452

Ardebili MS, Ghobadian B, Najafi G, Chegeni A (2011) Biodiesel production potential from edible oil seeds in Iran. Renew Sust Energ Rev 15:3041–3044

Ashnani MHM, Johari A, Hashim H, Hasani E (2014) A source of renewable energy in Malaysia, why biodiesel? Renew Sust Energ Rev 35:244–257

Atapour M, Kariminia H, Moslehabadi PM (2014) Optimization of biodiesel production by alkali-catalyzed transesterification of used frying oil. Process Saf Environ 92:179–185

Azadi P, Malina R, Barrett SRH, Kraft M (2017) The evolution of the biofuel science. Renew Sust Energ Rev 76:1479–1484

Bacenetti J, Restuccia A, Schillaci G, Failla S (2017) Biodiesel production from unconventional oilseed crops (*Linum usitatissimum* L. and *Camelina sativa* L.) in Mediterranean conditions: Environmental sustainability assessment. Renew Energ 112:444–456

Bahadar A, Khan MB (2013) Progress in energy from microalgae: a review. Renew Sust Energ Rev 27:128–148

Baskar G, Aiswarya R (2016) Trends in catalytic production of biodiesel from various feedstocks. Renew Sust Energ Rev 57:496–504

Batista AD, Amais RS, Rocha FRP (2016) Liquid-liquid microextraction in sequential injection analysis for the direct spectrophotometric determination of acid number in biodiesel. Microchem J 124:55–59

Bayrakci AG, Koçar G (2012) Utilization of renewable energies in Turkey's agriculture. Renew Sust Energ Rev 16:618–633

Bhattacharya S, Kabir KB, Hein K (2013) Dimethyl ether synthesis from Victorian brown coal through gasification—current status, and research and development needs. Prog Energy Combust Sci 39:577–605

Biodiesel (2017) http://www.anp.gov.br/wwwanp/biocombustiveis/biodiesel. Accessed 29 Jul 2017

Boletim Mensal do Biodiesel (2017) http://www.anp.gov.br/wwwanp/publicacoes/boletins-anp/2386-boletim-mensal-do-biodiesel. Accessed 29 Jul 2017

BRASIL. Lei nº 13.263, de 23 de março de 2016. Dispõe sobre os percentuais de adição de biodiesel ao óleo diesel comercializado no território nacional

Budžaki S, Miljić G, Tišma M, Sundaram S, Hessel V (2017) Is there a future for a enzymatic biodiesel industrial production in microreactors? Appl Energ 201:124–134

Bueno AV, Pereira MPB, de Pontes JVO, De Luna FMT, Cavalcante CL Jr (2017) Performance and emissions characteristics of castor oil biodiesel fuel blends. Appl Therm Eng 125:559–566

Campanelli P, Banchero M, Manna L (2010) Synthesis of biodiesel from edible, non-edible and waste cooking oils via supercritical methyl acetate transesterification. Fuel 89:3675–3682

Canam T, Li X, Holowachuk J, Yu M, Xia J, Mandal R, Krishnamurthy R, Bouatra S, Sinelnikov I, Yu B, Grenkow L, Wishart DS, Steppuhn H, Falk KC, Dumonceaux TJ, Gruber MY (2013) Differential metabolite profiles and salinity tolerance between two genetically related brown-seeded and yellow-seeded *Brassica carinata* lines. Plant Sci 198:17–26

Carlsson AS, Zhu L, Andersson M, Hofvander P (2014) Platform crops amenable to genetic engineering—a requirement for successful production of bio-industrial oils through genetic engineering. Biocatal Agric Biotechnol 3:58–64

Castanheira ÉG, Grisoli R, Freire F, Pecora V, Coelho ST (2014) Environmental sustainability of biodiesel in Brazil. Energ Policy 65:680–691

Chakraborty R, Bepari S, Banerjee A (2011) Application of calcined waste fish (*Labeo rohita*) scale as low-cost heterogeneous catalyst for biodiesel synthesis. Bioresource Technol 102:3610–3618

Chaves ES, Dos Santos EJ, Araujo RGO, Oliveira JV, Frescura VLA, Curtius AJ (2010) Metals and phosphorus determination in vegetable seeds used in the production of biodiesel by ICP OES and ICP-MS. Microchem J 96:71–76

Chen G, Tao J, Liu C, Yan B, Li W, Li X (2017) Hydrogen production via acetic acid steam reforming: a critical review on catalysts. Renew Sust Energ Rev 79:1091–1098

Chhikara S, Dutta I, Paulose B, Jaiwal PK, Dhankher OP (2012) Development of an *Agrobacterium*-mediated stable transformation method for industrial oilseed crop *Crambe abyssinica* 'BelAnn'. Ind Crops Prod 37:457–465

Corach J, Sorichetti PA, Romano SD (2016) Permittivity of biodiesel-rich blends with fossil diesel fuel: application to biodiesel content estimation. Fuel 177:268–273

da Costa MBR, Nicolau A, Guzatto R, Angeloni LM, Samios D (2016) Using biodiesel as a green solvent in the polymerization reactions: the attempt to separate the biodiesel from the polymer by thermal treatment. Polym Bull 74:2365–2378

da César AS, Werderits DE, de Saraiva GLO, da Guabiroba RCS (2017) The potential of waste cooking oil as supply for the Brazilian biodiesel chain. Renew Sust Energ Rev 72:246–253

Da Silva TRB, de Reis ACS, de Maciel CDG (2012) Relationship between chlorophyll meter readings and total N in crambe leaves as affected by nitrogen topdressing. Ind Crops Prod 39:135–138

De Marins AC, Reichert JM, Secco D, Rosa HA (2018) Crambe grain yield and oil content affected by spatial variability in soil physical properties. Renew Sust Energ Rev 81:464–472

De Oliveira FC, Coelho ST (2017) History, evolution and environmental impact of biodiesel in Brazil: a review. Renew Sust Energ Rev 75:168–179

Delparish A, Avci AK (2016) Intensified catalytic reactors for Fischer-Tropsch synthesis and for reforming of renewable fuels to hydrogen and synthesis gas. Fuel Process Technol 151:72–100

Demirbas A (2009) Progress and recent trends in biodiesel fuels. Energy Convers Manage 50: 14–34

Dhamodaran G, Krishnan R, Pochareddy YK, Pyarelal HM, Sivasubramanian H, Ganeshram AK (2017) A comparative study of combustion, emission, and performance characteristics of rice-bran-, neem-, and cottonseed-oil biodiesels with varying degree of unsaturation. Fuel 187:296–305

Dos Santos JI, Da Silva TRB, Rogério F, Santos RF, Secco D (2013) Yield response in crambe to potassium fertilizer. Ind Crops Prod 43:297–300

Dos Santos LK, Hatanaka RR, De Oliveira JE, Flumignan DL (2017) Experimental factorial design on hydroesterification of waste cooking oil by subcritical conditions for biodiesel production. Renew Energ 114:574–580

Dwivedi D, Dwivedi M, Malvíya S, Singh V (2017) Evaluation of wound healing, anti-microbial and antioxidant potential of *Pongamia pinnata* in wistar rats. J Tradit Complement Med 7: 79–85

Dwivedi G, Sharma MP (2015) Effect of metal on stability and cold flow property of pongamia biodiesel. Mat Today Proc 2:1421–1426

Eryilmaz T, Yesilyurt MK, Cesur C, Gokdogan O (2016) Biodiesel production potential from oil seeds in Turkey. Renew Sust Energ Rev 58:842–851

Espinoza A, Bautista S, Narváez PC, Alfaro M, Camargo M (2017) Sustainability assessment to support governamental biodiesel policy in Colombia: a system dynamics model. J Clean Prod 141:1145–1163

Faried M, Samer M, Abdelsalam E, Yousef RS, Attia YA, Ali AS (2017) Biodiesel production from microalgae: processes, technologies and recent advancements. Renew Sust Energ Rev 79:893–913

Farobie O, Matsumura Y (2017) Continuous production of biodiesel under supercritical methyl acetate conditions: Experimental investigation and kinetic model. Bioresource Technol 241:720–725

Fernández-Álvarez P, Vila J, Garrido JM, Grifoll M, Feijoo G, Lema JM (2007) Evaluation of biodiesel as bioremediation agent for the treatment of the shore affected by the heavy oil spill of the Prestige. J Hazard Mater 147:914–922

Ferreira CC, Costa LM, Barbeira PJS (2015) Methyl oleate as matrix simulacrum for the simultaneous determination of metals in biodiesel samples by flame atomic emission spectroscopy. Talanta 138:8–14

German L, Goetz A, Searchinger T, Oliveira G de LT, Tomei J, Hunsberger C, Weigelt J (2017) *Sine Qua Nons* of sustainable biofuels: Distilling implications of under-performance for national biofuel programs. Energ Policy 108:806–817

Goldemberg J, Coelho ST, Nastari PM, Lucon O (2004) Ethanol learning curve-the Brazilian experience. Biomass Bioenerg 26:301–304

Gonca G, Dobrucali E (2016) Theoretical and experimental study on the performance of a diesel engine fueled with diesel-biodiesel blends. Renew Energ 93:658–666

Gülüm M, Bilgin A (2015) Density, flash point and heating value variations of corn oil biodiesel-diesel fuel blends. Fuel Process Technol 134:456–464

Helwani Z, Othman MR, Aziz N, Kim J, Fernando WJN (2009) Solid heterogeneous catalysts for transesterification of triglycerides with methanol: a review. Appl Catal A-Gen 363:1–10

Hilioti Z, Michailof CM, Valasiadis D, Iliopoulou EF, Koidou V, Lappas AA (2017) Characterization of castor plant-derived biochars and their effects as soil amendments on seedlings. Biomass Bioenergy 105:96–106

Huang Y, Wu J (2008) Analysis of biodiesel promotion in Taiwan. Renew Sust Energ Rev 12:1176–1186

Hunsberger C, German L, Goetz A (2017) "Unbundling" the biofuel promise: querying the ability of liquid biofuels to deliver on socio-economic policy expectations. Energ Policy 108:791–805

Indrawan N, Thapa S, Rahman SF, Park J, Park S, Wijaya ME, Gobikrishnan S, Purwanto WW, Park D (2017) Palm biodiesel prospect in the Indonesian power sector. Environ Technol Innov 7:110–127

Ítavo LCV, De Souza ADV, Fávaro SP, Ítavo CCBF, Petit HV, Dias AM, Morais MG, Coelho RG, Reis FA, Costa JAA, Roscoe R (2016) Intake, digestibility, performance, carcass characteristics and meat quality of lambs fed different levels of crambe meal in the diet. Anim Feed Sci Technol 216:40–48

Khan SA, Rashmi Hussain MZ, Prasad S, Banerjee UC (2009) Prospects of biodiesel production from microalgae in India. Renew Sust Energ Rev 13:2361–2372

Korn MGA, Santos DCMB, Guida MAB, Barbosa IS, Passos MLC, Saraiva MLMFS, Lima JLFC (2010) Evaluation of digestion procedures for simultaneous determination of Ca, P, Mg, K and Na in biodiesel by inductively coupled plasma optical emission spectrometry. J Brazil Chem Soc 21:2278–2284

Kumar P, Srivastava VC, Jha MK (2016) *Jatropha curcas* phytotomy and applications: Development as a potential biofuel plant through biotechnological advancements. Renew Sust Energ Rev 59:818–838

Lamers P, McCormick K, Hilbert JA (2008) The emerging liquid biofuel market in Argentina: implications for domestic demand and international trade. Energ Policy 36:1479–1490

Lara-Fioreze ACC, Tomaz CA, Fioreze SL, Pilon C, Zanotto MD (2013) Genetic diversity among progenies of *Crambe abyssinica* Hochst for seed traits. Ind Crops Prod 50:771–775

Laviola BG, Rodrigues EV, Teodoro PE, de Peixoto LA, Bhering LL (2017) Biometric and biotechnology strategies in Jatropha genetic breeding for biodiesel production. Renew Sust Energ Rev 76:894–904

Lee J, Kim J, Ok YS, Kwon EE (2017) Rapid biodiesel synthesis from waste pepper seeds without lipid isolation step. Bioresource Technol 239:17–20

Leung DYC, Wu X, Leung MKH (2010) A review on biodiesel production using catalyzed transesterification. Appl Energ 87:1083–1095

Li X, Gao P, Gjetvaj B, Westcott N, Gruber MY (2009) Analysis of the metabolome and transcriptome of *Brassica carinata* seedlings after lithium chloride exposure. Plant Sci 177: 68–80

Lim BY, Shamsudin R, Baharudin BTHT, Yunus R (2016) Performance evaluation and CFD multiphase modeling for multistage jatropha fruit shelling machine. Ind Crops Prod 85: 125–138

Liu Y, Tu Q, Knothe G, Lu M (2017) Direct transesterification of spent coffee grounds for biodiesel production. Fuel 199:157–161

Lobo FA, Goveia D, De Oliveira AP, Pereira-Filho ER, Fraceto LF, Dias Filho NL, Rosa AH (2009) Comparison of the univariate and multivariate methods in the optimization of experimental conditions for determining Cu, Pb, Ni and Cd in biodiesel by GFAAS. Fuel 88:1907–1914

Lyra FH, Carneiro MTWD, Brandão GP, Pessoa HM, De Castro EV (2010) Determination of Na, K, Ca and Mg in biodiesel samples by flame atomic absorption spectrometry (F AAS) using microemulsion as sample preparation. Microchem J 96:180–185

Malaikozhundan B, Vaseeharan B, Vijayakumar S, Pandiselvi K, Kalanjiam R, Murugan K, Benelli G (2017) Biological therapeutics of *Pongamia pinnata* coated zinc oxide nanoparticles against clinically important pathogenic bacteria, fungi and MCF-7 breast cancer cells. Microb Pathog 104:268–277

Manique MC, Faccini CS, Onorevoli B, Benvenutti EV, Caramão EB (2012) Rice husk ash as an adsorbent for purifying biodiesel from waste frying oil. Fuel 92:56–61

Marillia E, Francis T, Falk KC, Smith M, Taylor DC (2014) Palliser's promise: *Brassica carinata*, an emerging western Canadian crop for delivery of new bio-industrial oil feedstocks. Biocatal Agric Biotechnol 3:65–74

Mariprasath T, Kirubakaran V (2016) A critical review on the characteristics of alternating liquid dielectrics and feasibility study on pongamia pinnata oil as liquid dielectrics. Renew Sust Energ Rev 65:784–799

Martínez G, Sánchez N, Encinar JM, González JF (2014) Fuel properties of biodiesel from vegetable oils and oil mixtures. Influence of methyl esters distribution. Biomass Bioenergy 63:22–32

Meloni D, Perra D, Monaci R, Cutrufello MG, Rombi E, Ferino I (2016) Transesterification of *Jatropha curcas* oil and soybean oil on Al-SBA-15 catalysts. Appl Catal B 184:163–173

Mishra S, Anand K, Santhosh S, Mehta PS (2017) Comparison of biodiesel fuel behavior in a heavy duty turbocharged and a light duty naturally aspirated engine. Appl Energ 202:459–470

Montes JM, Melchinger AE (2016) Domestication and Breeding of *Jatropha curcas* L. Trends Plant Sci 21:1045–1057

Mudalkar S, Sreeharsha RV, Reddy AR (2016) A novel aldo-keto reductase from *Jatropha curcas* L. (JcAKR) plays a crucial role in the detoxification of methylglyoxal, a potente electrophile. J Plant Physiol 195:39–49

Nalgundwar A, Paul B, Sharma SK (2016) Comparison of performance and emissions characteristics of DI CI engine fueled with dual biodiesel blends of palm and jatropha. Fuel 173:172–179

Naresh B, Reddy MS, Vijayalakshmi P, Reddy V, Devi P (2012) Physico-chemical screening of accessions of *Jatropha curcas* for biodiesel production. Biomass Bioenergy 40:155–161

Naylor RL, Higgins MM (2017) The political economy of biodiesel in an era of low oil prices. Renew Sust Energ Rev 77:695–705

Negm NA, Sayed GH, Habib OI, Yehia FZ, Mohamed EA (2017) Heterogeneous catalytic transformation of vegetable oils into biodiesel in one-step reaction using super acidic sulfonated modified mica catalyst. J Mol Liq 237:38–45

Nguyen HC, Liang S, Doan TT, Su C, Yang P (2017) Lipase-catalyzed synthesis of biodiesel from black soldier fly (*Hermetica illucens*): Optimization by using response surface methodology. Energ Convers Manage 145:335–342

No DS, Zhao T, Kim BH, Choi H, Kim I (2013) Enrichment of erucic acid from crambe oil in a recirculated packed bed reactor via lipase-catalyzed ethanolysis. J Mol Catal B Enzym 87:6–10

Nogueira T, Do Lago CL (2011) Determination of Ca, K, Mg, Na, sulfate, phosphate, formate, acetate, propionate, and glycerol in biodiesel by capillary electrophoresis with capacitively coupled contactless conductivity detection. Microchem J 99:267–272

Oliveira RS, Machado PMA, Ramalho HF, Rangel ET, Suarez PAZ (2017) Acylation of epoxidized soybean biodiesel catalyzed by $SnO/Al_2O_3$ and evaluation of physical chemical and biologic activity of the product. Ind Crop Prod 104:201–209

Omonov TS, Kharraz E, Curtis JM (2017) Camelina (*Camelina Sativa*) oil polyols as an alternative to Castor oil. Ind Crop Prod 107:378–385

Orozco FDA, Kovachev N, Pastor MÁA, Domini CE, Band BSF, Hernández AC (2014) Analysis of metals and phosphorous in biodiesel B100 from different feedstock using a Flow Blurring® multinebulizer in inductively coupled plasma-optical emission spectrometry. Anal Chim Acta 827:15–21

Pandit PR, Fulekar MH (2017) Egg shell waste as heterogeneous nanocatalyst for biodiesel production: optimized by response surface methodology. J Environ Manage 198:319–329

Panorama Internacional Petróleo (2017) http://www.anp.gov.br/wwwanp/images/publicacoes/anuario-estatistico/2017/Textos/Secao1.pdf. Accessed 17 Sept 2017

Pereira FM, Zimpeck RC, Brum DM, Cassella RJ (2013) Novel extraction induced by emulsion breaking as a tool for the determination of trace concentrations of Cu, Mn and Ni in biodiesel by electrothermal atomic absorption spectrometry. Talanta 117:32–38

Pitol C, Broch DL, Roscoe R (2010) Tecnologia e Produção: Crambe. 1ª ed. Maracaju: Fundação MS 60 p

Pousa GPAG, Santos ALF, Suarez PAZ (2007) History and policy of biodiesel in Brazil. Energ Policy 35:5393–5398

Quadros DPC, Rau M, Idrees M, Chaves ES, Curtius AJ, Borges DLG (2011) A simple and fast procedure for the determination of Al, Cu, Fe and Mn in biodiesel using high-resolution continuum source electrothermal atomic absorption spectrometry. Spectrochim Acta B 66: 373–377

Quintero JA, Felix ER, Rincón LE, Crisspín M, Baca JF, Khwaja Y, Cardona CA (2012) Social and Techno-economical analysis of biodiesel production in Peru. Energ Policy 43:427–435

Radhakumari M, Taha M, Shahsavari E, Bhargava SK, Satyavathi B, Ball AS (2017) *Pongamia pinnata* seed residue—a low cost inedible resource for on-site/in-house lignocellulases and sustainable ethanol production. Renew Energ 103:682–687

Rajaeifar MA, Abdi R, Tabatabaei M (2017) Expanded polystyrene waste application for improving biodiesel environmental performance parameters from life cycle assessment point of view. Renew Sust Energ Rev 74:278–298

Ramos MJ, Fernández CM, Casas A, Rodríguez L, Pérez Á (2009) Influence of fatty acid composition of raw materials on biodiesel properties. Bioresource Technol 100:261–268

Ranganathan SV, Narasimhan SL, Muthukumar K (2008) An overview of enzymatic production of biodiesel. Bioresource Technol 99:3975–3981

Rattanapoltee P, Kaewkannetra P (2014) Cultivation of microalga, *Chlorella vulgaris* under different auto-hetero-mixo trophic growths as a raw material during biodiesel production and cost evaluation. Energy 78:4–8

REN21 (2017) Renewables 2017 global status report. http://www.ren21.net/gsr-2017/. Accessed 14 June 2017

Rico JAP, Sauer IL (2015) A review of Brazilian biodiesel experiences. Renew Sust Energ Rev 45:513–529

Rodionova MV, Poudyal RS, Tiwari I, Voloshin RA, Zharmukhamedov SK, Nam HG, Zayadan BK, Bruce BD, Hou HJM, Allakhverdiev SI (2017) Biofuel production: challenges and opportunities. Int J Hydrogen Energy 42:8450–8461

Rodrigues J, Miranda I, Gominho J, Vasconcelos M, Barradas G, Pereira H, Bianchi-de-Aguiar F, Ferreira-Dias S (2016) Modeling and optimization of laboratory-scale conditioning of *Jatropha curcas* L. seeds for oil expression. Ind Crops Prod 83:614–619

Rogério F, Da Silva TRB, Dos Santos JI, Poletine JP (2013) Phosphorus fertilization influences grain yield and oil content in crambe. Ind Crops Prod 41:266–268

Roveda AC, Comin M, Caires ARL, Ferreira VS, Trindade MAG (2016) Thermal stability enhancement of biodiesel induced by a synergistic effect between conventional antioxidants and an alternative additive. Energy 109:260–265

Saba N, Jawaid M, Hakeem KR, Paridah MT, Khalina A, Alothman OY (2015) Potential of bioenergy production from industrial kenaf (*Hibiscus cannabinus* L. based on Malaysian perspective. Renew Sust Energ Rev 42:446–459

Sahu B, Sahu AK, Thomas V, Naithani SC (2017) Reactive oxygen species, lipid peroxidation, protein oxidation and antioxidative enzymes in dehydrating Karanja (*Pongamia pinnata*) seeds during storage. S Afr J Bot 112:383–390

Santos KA, Bariccatti RA, Cardozo-Filho L, Schneider R, Palú F, Da Silva C, Da Silva EA (2015) Extraction of crambe seed oil using subcritical propane: Kinetics, characterization and modeling. J Supercrit Fluids 104:54–61

Seepaul R, George S, Wright DL (2016) Comparative response of *Brassica carinata* and *B. napus* vegetative growth, development and photosynthesis to nitrogen nutrition. Ind Crops Prod 94:872–883

Selo Combustível Social (2015) http://www.mda.gov.br/sitemda/noticias/selo-combust%C3%ADvel-social. Accessed 29 Jul 2017

Shah A, Gor K, Tripathi S, Nair A, Veerappapillai S (2016) Aplication of biodiesel in metalworking fluids as a substitute for petroleum solvent. Int J Pharm Sci Rev Res 38:106–108

Shimamoto GG, Tubino M (2016) Alternative methods to quantify biodiesel in standard diesel-biodiesel blends and sample adulterated with vegetable oil through UV-Visible spectroscopy. Fuel 186:199–203

Shu Q, Gao J, Nawaz Z, Liao Y, Wang D, Wang J (2010) Synthesis of biodiesel from waste vegetable oil with large amounts of free fatty acids using a carbon-based solid acid catalyst. Appl Energ 87:2589–2596
Sikarwar VS, Zhao M, Fennell PS, Shah N, Anthony EJ (2017) Progress in biofuel production from gasification. Prog Energy Combust Sci 61:189–248
Silva LMA, Andrade FD, Filho EGA, Monteiro MR, De Azevedo ER, Venâncio T (2016a) NMR investigation of commercial carbon black filled vulcanized natural rubber exposed to petrodiesel/biodiesel mixtures. Fuel 186:50–57
Silva WC, Castro MPP, Perez VH, Machado FA, Mota L, Sthel MS (2016b) Thermal degradation of ethanolic biodiesel: Physicochemical and thermal properties evaluation. Energy 114:1093–1099
Singh AS, Kumari S, Modi AR, Gajera BB, Narayanan S, Kumar N (2015) Role of conventional and biotechnological approaches in genetic improvement of castor (*Ricinus communis* L.). Ind Crop Prod 74:55–62
Singh L, Wahid ZA (2015) Methods for enhancing bio-hydrogen production from biological process: a review. J Ind Eng Chem 21:70–80
Squissato AL, Lima AF, Almeida ES, Pasquini D, Richter EM, Munoz RAA (2017) Eucalyptus pulp as an adsorbent for metal removal from biodiesel. Ind Crop Prod 95:1–5
Stamigna C, Chiaretti D, Chiaretti E, Prosini PP (2012) Oil and furfural recovery from *Brassica carinata*. Biomass Bioenerg 39:478–483
Su Y, Song K, Zhang P, Su Y, Cheng J, Chen X (2017) Progress of microalgae biofuel's commercialization. Renew Sust Energ Rev 74:402–411
Swain PK, Das LM, Naik SN (2011) Biomass to liquid: A prospective challenge to research and development in 21st century. Renew Sust Energ Rev 15:4917–4933
Tan KT, Lee KT, Mohamed AR (2010) Effects of free fatty acids, water content and co-solvent on biodiesel production by supercritical methanol reaction. J Supercrit Fluid 53:88–91
Tarikahya-Hacioğlu B (2016) Molecular diversity of the wild *Crambe* (Brassicaceae) taxa in Turkey detected by inter-simple sequence repeats (ISSRs). Ind Crops Prod 80:214–219
Tavares GR, Massa TB, Gonçalves JE, Da Silva C, Dos Santos WD (2017) Assessment of ultrasound-assisted extraction of crambe seed oil for biodiesel synthesis by in situ interesterification. Renew Energ 111:659–665
Thangaraja J, Anand K, Mehta PS (2016) Biodiesel $NO_x$ penalty and control measures—a review. Renew Sust Energ Rev 61:1–24
Tongroon M, Suebwong A, Kananont M, Aunchaisri J, Chollacoop N (2017) High quality jatropha biodiesel (H-FAME) and its application in a common rail diesel engine. Renew Energ 113:660–668
Torrentes-Espinoza G, Miranda BC, Vega-Baudrit J, Mata-Segreda JF (2017) Castor oil (*Ricinus communis*) supercritical methanolysis. Energy 140:426–435
Ullah K, Ahmad M, Sofia Sharma VK, Lu P, Harvey A, Zafar M, Sultana S (2015) Assessing the potential of algal biomass opportunities for bioenergy industry: a review. Fuel 143:414–423
Vicente G, Carrero A, Rodríguez R, Del Peso GL (2017) Heterogeneous-catalyzed direct transformation of microalga biomass into biodiesel-grade FAMEs. Fuel 200:590–598
Vicente G, Martínez M, Aracil J (2004) Integrated biodiesel production: a comparison of different homogeneous catalysts systems. Bioresour Technol 92:297–305
Wang X, Qin X, Li D, Yang B, Wang Y (2017) One-step synthesis of high-yield biodiesel from waste cooking oils by a novel and highly methanol-tolerant immobilized lipase. Bioresource Technol 235:18–24
Wazilewski WT, Bariccatti RA, Martins GI, Secco D, De Souza SNM, Rosa HA, Chaves LI (2013) Study of the methyl crambe (*Crambe abyssinica* Hochst) and soybean biodiesel oxidative stability. Ind Crops Prod 43:207–212
Xia A, Cheng J, Murphy JD (2016) Innovation in biological production and upgrading of methane and hydrogen for use as gaseous transport biofuel. Biotechnol Adv 34:451–472
Xiang Y, Xiang Y, Wang L (2017) Microwave radiation improves biodiesel yields from waste cooking oil in the presence of modified coal fly ash. J Taibah Univ Sci 11:1019–1029

Xin H, Yu P (2014) Detect changes in lipid-related structure of brown- and yellow-seeded *Brassica Carinata* seed during rumen fermentation in relation to basic chemical profile using ATr-FT/IR molecular spectroscopy with chemometrics. Spectrochim Acta A 133:811–817

Xiong W, Yu-hua Z, Cheng-chao L, Jing-ping H, Liang W, Yao C, Jin-lin L (2017) Performance of hierarchical ZSM-5 supported cobalt catalyst in the Fischer-Tropsch synthesis. J Fuel Chem Technol 45:950–955

Xu Y, Li G, Sun Z (2006) Development of biodiesel industry in China: Upon the terms of production and consumption. Renew Sust Energ Rev 54:318–330

Xue Y, Zhao Z, Xu G, Lian X, Yang C, Zhao W, Ma P, Lin H, Han S (2016) Effect of poly-alpha-olefin pour point depressant on cold flow properties of waste cooking oil biodiesel blends. Fuel 184:110–117

Yaakob Z, Mohammad M, Alherbawi M, Alam Z, Sopian K (2013) Overview of the production of biodiesel from waste cooking oil. Renew Sust Energ Rev 18:184–193

Yang L, Takase M, Zhang M, Zhao T, Wu X (2014) Potential non-edible oil feedstock for biodiesel production in Africa: a survey. Renew Sust Energ Rev 38:461–477

Yin X, Zhang X, Wan M, Duan X, You Q, Zhang J, Li S (2017) Intensification of biodiesel production using dual-frequency counter-current pulsed ultrasound. Ultrason Sonochem 37:136–143

Zhang W, Yan W, Zhang X, Coronado M (2012) Predicting the dynamic and kinematic viscosities of biodiesel-diesel blends using mid- and near-infrared spectroscopy. Appl Energ 98:122–127

Zhang X, Qian W, Zhang H, Sun Q, Ying W (2017) Effect of the operation parameters on the Fischer-Tropsch synthesis in fluidized bed reactors. Chin J Chem Eng In Press. https://doi.org/10.1016/j.cjche.2017.05.012

Zhang X, Yang Z, Liang J, Tang L, Chen F (2016) Detoxification of Jatropha curcas seed cake in solid-state fermentation of newly isolated endophytic strain and nutrition assessment for its potential utilizations. Int Biodeterior Biodegrad 109:202–210

Živković SB, Veljković MV, Banković-Ilić IB, Krstić IM, Konstantinović SS, Ilić SB, Avramović JM, Stamenković OS, Veljković VB (2017) Technological, technical, economic, environmental, social, human health risk, toxicological and policy considerations of biodiesel production and use. Renew Sust Energ Rev 79:222–247

# Chapter 2
# Employment of Alternative Raw Materials for Biodiesel Synthesis

**Bruna Silveira Pacheco, Caroline Carapina da Silva, Samantha Coelho de Freitas, Lucas Moraes Berneira, Vinícius Lenz da Silva, Kathleen Winkel, Letícia Braatz Ferreira and Claudio Martin Pereira de Pereira**

**Abstract** The search for renewable fuels has increased in recent times, since petroleum-derived fuels are depleting supplies and contributing to accumulation of carbon dioxide in the environment. In this way, biodiesel presents economic, social, and environmental advantages as a renewable biofuel. This chapter aims to inform the reader about different raw materials that can be used to synthesize biodiesel. Moreover, the main physicochemical parameters are presented based on the standard agencies around the world. In general, the raw materials (algae, castor oil, soybean, waste frying oil, and *Chrysomya albiceps* larvae) represented satisfactory sources to produce biodiesel.

**Keywords** Biodiesel · Raw materials · Vegetable oil · Biodiesel synthesis
Renewable fuels

---

B. S. Pacheco
Post Graduate Program in Biotechnology, Laboratory of Lipidomics and Bioorganic, Center of Chemical, Pharmaceutical and Food Sciences, Federal University of Pelotas, Pelotas, Rio Grande do Sul 96001-970, Brazil

C. C. da Silva
Post Graduate Program in Biochemistry and Bioprospection, Laboratory of Lipidomics and Bioorganic, Center of Chemical, Pharmaceutical and Food Sciences, Federal University of Pelotas, Pelotas, Rio Grande do Sul 96001-970, Brazil

S. C. de Freitas · L. M. Berneira · V. L. da Silva · K. Winkel
L. B. Ferreira · C. M. P. de Pereira (✉)
Laboratory of Lipidomics and Bioorganic, Center of Chemical, Pharmaceutical and Food Sciences, Federal University of Pelotas, Pelotas, Rio Grande do Sul 96001-970, Brazil
e-mail: claudiochemistry@gmail.com; lahbbioufpel@gmail.com



M. Trindade (ed.), *Increased Biodiesel Efficiency*, Green Energy and Technology,
https://doi.org/10.1007/978-3-319-73552-8_2

## 2.1 Biodiesel

Biodiesel is a biofuel characterized by a mixture of fatty acid alkyl esters. The most common way to produce biodiesel is by transesterification. In this reaction, triglycerides (TG), the main components of vegetable oils, react with an alcohol to produce fatty acid monoalkyl esters and glycerol. It can be produced from diverse raw materials such as vegetable oils (Vieira et al. 2016), microorganisms (Hobuss et al. 2011), and waste materials (Boey et al. 2012). Nowadays, this alternative energy source is receiving more attention due to the high rate of pollution, health damage, and environmental impact caused by fossil fuels. Moreover, most of petroleum and its derivatives' reserves are heading to depletion and, thus, enlarging the importance of biodiesel in the near future (Ramos et al. 2009; Bazooyar et al. 2011; Sajjadi et al. 2016).

Historically, the first use of vegetable oils as fuel that occurred in early twentieth century when France was interested in providing local fuel aiming for its distant colonies becomes energetically independent. From this first incentive to the end of World War II (WWII), there were several reports in the literature about the potential use of vegetable oils as fuels in a way that they were also related to energy independence. However, after WWII its use as fuel declined and it was largely substituted by fossil fuels. Nevertheless, vegetable oils became again a promising source of combustible mainly because of the energy crisis that emerged in the late 1970s. Nowadays, along with an environmental purpose, these oils are returning to gain prominence mainly as biodiesel (Knothe and Razon 2017).

It is widely known that energy demand will increase dramatically in the next years, due to factors such as a population growth and economic development. However, the main source of power in the whole planet is still based on fossil fuels that are a non-renewable source. For instance, in 2015 almost 86% of the energy consumed in the world came from petroleum-based sources. That has led to an consumption of 11 billion tons yearly of fossil fuels in 2015, which contributed to depletion of natural resources, global warming, degraded environmental quality, besides economic and human health crisis (Sajjadi et al. 2016). In this order, energy sources which can be renewable and safe to the environment are important to ensure good quality of life to the future generations (Bazooyar et al. 2011; Oliveira et al. 2013; Knothe and Razon 2017). In this concept, Table 2.1 compares the gases emitted by diesel, a 20% blend of biodiesel in diesel (B20) and pure biodiesel (B100).

**Table 2.1** Diesel, B20, and B100 engine emission on maximum speed

| | Diesel | B20 | B100[a] |
|---|---|---|---|
| Carbon monoxide (ppm) | 1292 | 1172 | 456 |
| Carbon dioxide (%) | 9.38 | 8.86 | 7.84 |
| Nitrogen oxide (ppm) | 1275 | 1251 | 1180 |
| Nitrogen dioxide (mg/Nm$^3$) | 2873 | 2716 | 2214 |
| Oxygen (%) | 8.06 | 8.86 | 14.4 |

Adapted from (Ilkiliç 2010)
[a]Properties in accordance with ASTM specifications

**Table 2.2** Diesel and rice bran biodiesel (B100) physicochemical properties

| Properties | Diesel | B100[a] |
|---|---|---|
| Density 26 °C (kg $L^{-1}$) | 0.84 | 0.88 |
| Viscosity 26 °C ($mm^2$ $s^{-1}$) | 3.20 | 4.30 |
| Calorific value (kJ $kg^{-1}$) | 42.90 | 40.76 |
| Flash point (°C) | 58 | 148 |
| Cetane number | 48 | 51 |
| Sulfur value (ppm) | 0.05 | <0.01 |
| Acidic index | 0.22 | 0.13 |

Adapted from (Ilkiliç 2010)
[a]Properties in accordance with ASTM specifications

Biodiesel is known for being a biodegradable, non-toxic, and renewable source of combustible compared to fossil fuels. Moreover, it has low or no sulfur content, inherent lubricity, and high flash point—features that make it a very promising biofuel. However, biodiesel faces several challenges in order to be fully adopted as a common combustible, such as higher cost compared to fossil fuels and parameters as oxidative stability, cold flow properties, and viscosity (Ramos et al. 2009; Sajjadi et al. 2016). In this context, future research is needed for accomplishing these parameters set by standard agencies around the world and, thus, makes biodiesel a feasible source of energy (Knothe and Razon 2017). Table 2.2 exposes some parameters seen in pure biodiesel (B100) and diesel.

According to the literature, worldwide production of biodiesel reaches approximately 29.7 billion liters. From this amount, 4.7 billion liters were produced in the USA, followed by Brazil (3.4 billion liters), Indonesia (3.1 billion liters), and Argentina (2.9 billion liters). The European Union accounted for 11.6 billion liters altogether. The production of biodiesel is increasing year after year and however still stumbles in feedstock availability and agricultural land even with numerous subsidies and regulatory efforts facilitating its production and use (Knothe and Razon 2017).

### 2.1.1 General Procedure for Biodiesel Synthesis

Early studies on vegetable oils as a source of energy used raw vegetable oils; however, high viscosity, density, iodine values, and poor volatility of these oils led to issues such as gumming and formation of carbon deposits in the engines. In order to improve these parameters, several methods have been developed, such as microemulsification, pyrolysis, hydrodeoxygenation, and transesterification (Bazooyar et al. 2011; Sajjadi et al. 2016).

Transesterification has become the most employed methodology since it is economically advantageous compared to the other procedures. This method converts almost 98% of triglycerides into alkyl esters (biodiesel) through the reaction of triacylglycerol with a short-chain alcohol, being the main products of this reaction alkyl esters and glycerol as by-product. Nowadays, methanol is the most employed alcohol in transesterification, since it is cheaper and reacts more easily than other compounds from the same chemical class. Alkali catalysts, such as sodium and potassium hydroxide, are also more used than acid catalysts in this type of reaction, because alkali-catalyzed transesterification can be done at low temperatures and pressures, which decrease its production/commercial cost (Ramos et al. 2009; Bazooyar et al. 2011).

Biodiesel can be synthesized from several sources (i.e., there are more than 350 crops recognized as potential source for biodiesel production); however, soybean oil, rapeseed oil, palm oil, and sunflower seed oil dominate its production. Generally, biodiesel feedstock can be divided into four classes such as edible, non-edible, waste oils, and animal fats; therefore, more than 95% of the commercial biodiesel comes from edible oils. The transesterification reaction does not alter the composition of the acylglycerol; thus, the characteristics of biodiesel are defined by the fatty acid composition of the raw material. In this order, the chain length of the fatty acids is related to biodiesel parameters, such as density and viscosity due to fact that the unsaturation index is related to the oxidative stability of the biofuel (Ramos et al. 2009).

It is known that approximately 75% of the biodiesel cost comes from feedstock; thus, choosing a proper source is essential. In this context, non-edible and waste oils are eminent sources, since they are not used for human nutrition and can be obtained from several regions of the planet, which could reduce deforestation and the fuel/food debate (Ramos et al. 2009). Waste oils are promising sources to produce biodiesel, because the oil is reutilized instead of discarded into the environment, minimizing the pollutant release which could be released to the environment. However, there are still several barriers for its widely reutilization, due to the infrastructure and logistics needed for this type of operation. Moreover, waste oils must go through several steps of cleaning before transesterification, which could increase the production costs (Sajjadi et al. 2016).

Triglycerides are the main components of the raw oils from vegetable sources which are largely used for biodiesel production. In transesterification reaction, triglycerides are converted to diglyceride, monoglyceride, and finally converted to glycerol. Transesterification reaction to obtain biodiesel is represented in Scheme 2.1. Below are presented the main methodologies used for biodiesel synthesis.

#### 2.1.1.1 Alkaline Catalysis

Alkaline-catalyzed homogeneous transesterification is the most common and cheaper method employed to biodiesel production. Generally, sodium hydroxide (NaOH), sodium methoxide ($CH_3ONa$), potassium hydroxide (KOH), and potassium methoxide ($CH_3OK$) are the most common catalysts. These catalysts are more effective than the acids; thus, transesterification reaction occurs faster (Rashid and Anwar 2008; Ehsan and Chowdhury 2015).

To production of biodiesel by alkaline catalysis, potassium hydroxide (0.34 g) was dissolved in methanol (26 mL). This solution was diluted in 50 mL of selected

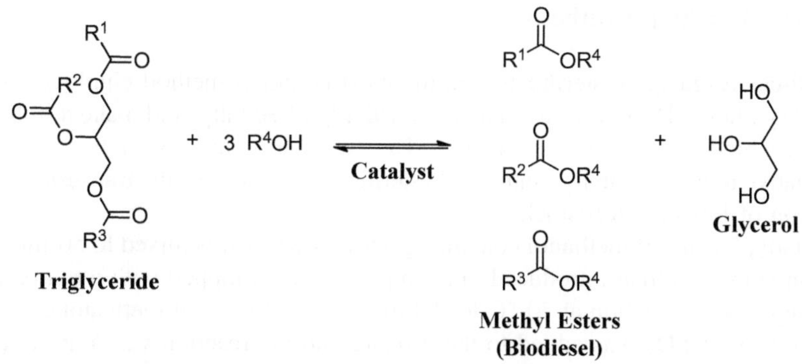


**Scheme 2.1** Transesterification process to biodiesel production

oil, and the resulted content was put under reflux and constant stirring during 1 h. Afterward, the products of the reaction were transferred to a separating funnel in order to separate biodiesel from glycerol. Biodiesel was washed with solution of chloridric acid (30 mL, 0.5% v/v) and subsequently with a saturated aqueous solution of sodium chloride (NaCl) (30 mL). Finally, anhydrous sodium sulfate ($Na_2SO_4$) was added to organic phase, the biodiesel was filtered, and the solvent was evaporated under vacuum.

#### 2.1.1.2 Acidic Catalysis

As a general rule, the use of acid catalysts produces lower quality and lower yield biodiesel than the alkali-catalyzed transesterification. However, when the raw material is rich in free fatty acids the acidic catalysts have showed very good yields. Based on that, acidic catalysis has being widely used to produce biodiesel from waste frying oil (Felizardo et al. 2006). The main disadvantage of acidic-catalyzed homogeneous reactions is that the process is about 4000 times slower than the alkali. Sulfuric acid ($H_2SO_4$), hydrofluoric acid (HF), and *p*-toluenesulfonic (*p*-TSA) acid are the most common acids used in biodiesel synthesis by acid catalysis (Lotero et al. 2005).

For the synthesis of biodiesel, firstly, 1.5 g of *p*-toluenesulfonic acid was dissolved in 40 mL of methanol. This solution was diluted in 50 mL of oil, and the resulted content was put under reflux and constant stirring during 2 h. After that, the products of the reaction were transferred to a separating funnel in order to separate biodiesel from glycerol. Sequentially, biodiesel was washed with 30 mL of distillated water and with 30 mL of ethyl ether. Finally, anhydrous sodium sulfate was added to this phase, the biodiesel was filtered, and the solvent was evaporated under vacuum.

#### 2.1.1.3 Two-Step Synthesis

Alkali-catalyzed transesterification is the most common method employed to biodiesel synthesis. However, raw materials with high free fatty acid make it difficult to produce biodiesel using this method. The two-step synthesis of biodiesel is an alternative to ensure the free fatty acid esterification as well as the transesterification reaction of low-cost feedstocks.

Firstly, 20 mL of methanol containing 0.5 g of KOH was mixed to 50 mL of oil. The mixture was then introduced in a simple reactor equipped with a reflux device and kept under agitation at 70 °C for 60 min. Then, 30 mL of methanol containing 0.75 mL of $H_2SO_4$ was added to the mixture and the reaction was kept under the same conditions for another hour. Soon after the transesterification reaction had ended and the mixture had cooldown, it was filtrated using a vacuum pump (to separate the salts formed when acid and base were mixed) and transferred to a separating funnel (to separate the glycerin). Lastly, the biodiesel was washed three times with 50 mL of water at 70 °C, taken to the rotary evaporator (to eliminate the volatiles), and dried by bubbling nitrogen ($N_2$) for 15 min under a 0.4 mL $min^{-1}$ flux (Oliveira et al. 2013).

### *2.1.2 Quality Parameters*

The analysis of biodiesel comprises several tests to measure the quality of the biofuel synthesized, so as the possible contaminants. Generally, these parameters are set by regulatory agencies. The most important of these organizations are American Society for Testing and Materials (ASTM) and European Standards (EN). However, other countries have their own legislation, such as Brazil (National Agency of Petroleum, Natural Gas and Biofuels—ANP). ASTM published the first quality parameters for pure biodiesel (B100) in 2002. Since then, some updates were made including several new methods to ensure the biodiesel quality (Alleman et al. 2013; Knothe and Razon 2017).

In 2006, almost 59% of biodiesel samples collected in the USA were out of the parameters set by ASTM. The analyzed biofuels failed in parameters such as flash point, oxidative stability, and glycerol content. In this context, methods that can determine the quality of biodiesel are extremely important, since low-quality fuels cause several issues in vehicles as well as in the environment (Alleman et al. 2013).

Regarding biodiesel quality, the following section presents the most important parameters and the main analytical methods performed to investigate such standards.

#### 2.1.2.1 Kinematic Viscosity

Reducing viscosity is one of the main reasons to the production of biodiesel by transesterification. This property is of extreme importance to the quality of biodiesel, since a high viscosity fuel leads to several engine problems, such as engine deposits and gum formation (Knothe and Steidley 2005). Generally, the viscosity of biodiesel increases as the fatty acid's chain length is longer and decreases as its unsaturation level increases. Following this line of thought, Crizel et al. (2016) made use of viscosity analysis as a potential tool to detect adulteration in biodiesel.

Saybolt viscometer is used as a standard test by ASTM. To this test, the biodiesel must be added to a preheated viscometer at 40 °C. Sequentially, biodiesel is run gradually under gravitational force while the time for its complete downfall is measured. According to ASTM, kinematic viscosity at 40 °C must be between 1.9 and 6 $mm^2\ s^{-1}$ while European standards set a range of 3.5–5 $mm^2\ s^{-1}$.

#### 2.1.2.2 Flash Point

This property is related to the minimum temperature which a volatile liquid releases a sufficient amount of vapor to ignite. Vegetable oils tend to have a greater value of flash point when compared to its respective alkyl esters. Residues from the synthesis, such as alcohol, can also decrease this parameter; thus, the flash point could be used to determine the residual alcohol in biodiesel (Boog et al. 2011). This property is important to comprehend storing conditions, transporting, and handling as well other circumstances of safety. According to ASTM, flash point must be greater than 130 °C while European standards accept a value greater than 120 °C (Alleman et al. 2013; Knothe and Razon 2017).

Experimentally, when the test starts, biodiesel is exposed to a flame that gradually increases its temperature. The lowest temperature at which the biofuel suffers ignition is recorded and the test is finalized. The flash point values determined for biodiesel are relatively higher than those of mineral diesel (approximately 55 °C).

#### 2.1.2.3 Iodine Value

This parameter measures the total unsaturation level of oil, and this value is used to assess the oxidative stability of the biodiesel. The test is related to the interaction between iodine and unsaturated fatty acids present in the biodiesel. According to European standards, iodine value must be lower than 120 g $\times$ 100 $g^{-1}$. There is no American legislation about this parameter (Knothe and Razon 2017).

Iodine value can be determined experimentally by the dissolution of biodiesel in cyclohexane and Wijs solution. The iodine value is found by titrating with sodium thiosulphate ($Na_2S_2O_3$) solution (Sandha and Swami 2009).

### 2.1.2.4 Saponification Value

This test is used as an indicator of average molecular weight. Generally, vegetable oil and its respective biodiesel tend to have similar saponification values (Endo et al. 2005). Since this property is only a relative measure to predict the chain length of all fatty acids in the oil sample. The USA and European standards do not regulate it. The saponification value is experimentally determined by the reaction of bio-diesel with a certain amount of potassium hydroxide (KOH) and subsequent titration with HCl. The saponification value is expressed as the amount of alkali (sodium or potassium hydroxide) required to complete saponification of 1.0 g of the oil. The reaction could be visualized in Scheme 2.2.

### 2.1.2.5 Thin-Layer Chromatography

Chromatography is an area of chemistry that is related to separation of a specific compound in a complex sample. In this context, it is an important tool to evaluate the progress of biodiesel synthesis through the fatty acid profile. Chromatography is not regulated by standard agencies and is performed as a complementary study to determine some parameters of this biofuel (Oliveira et al. 2013).

Thin-layer chromatography (TLC) is a type of chromatography which is based on different interactions between the sample and the stationary and mobile phases, in order to separate them. This procedure can be experimentally done through the dissolution of the biodiesel and application in the stationary phase of the TLC plate that will be placed in contact with stationary phase. After the effect of the capillary action, the compounds are separated and are prepared to be revealed by iodine vapors. An example comparing the vegetable oil with the respective biodiesel is shown in Fig. 2.1.

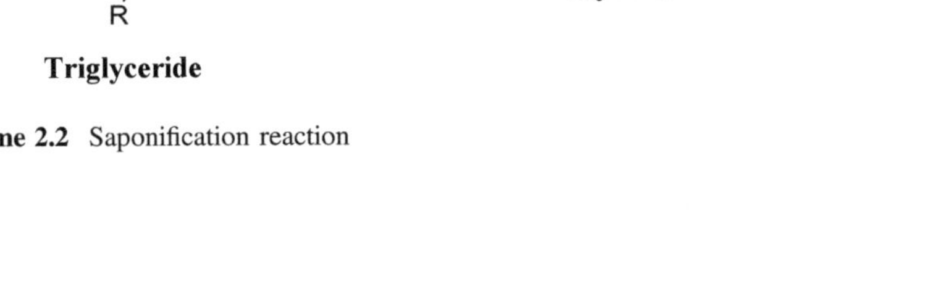


**Scheme 2.2** Saponification reaction

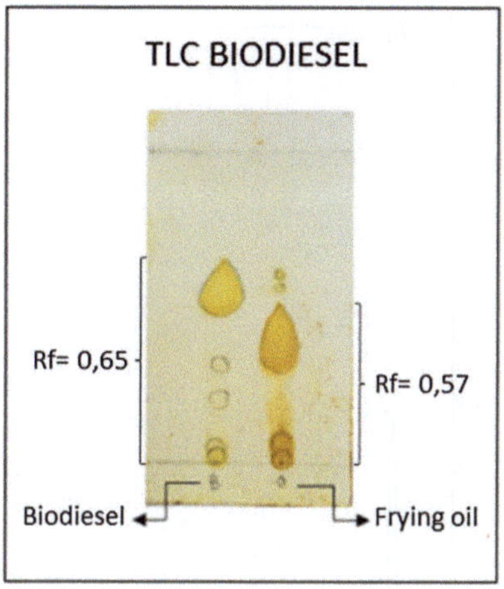


**Fig. 2.1** Thin-layer chromatography comparing waste frying oil and its biodiesel (Rf = Retention factor)

#### 2.1.2.6 Gas Chromatography

Many chromatographic methods have been employed for biodiesel analysis, and in this context, gas chromatography (GC) is a fundamental tool to determine biodiesel's fatty acid composition. This measure is an important feature since this profile is related to several other parameters analyzed which are fundamental to ensure biodiesel's quality, although there is no legislation in the USA or Europe regarding GC analysis (Alleman et al. 2013).

Experimentally, the fatty acid profile of biodiesel can be determined as fatty acid methyl esters (FAME) through GC analysis as described by Hobuss and co-authors (2011). Fatty acid profile confirmation is obtained by comparing with a standard FAME Mix. Table 2.3 presents GC conditions.

**Table 2.3** GC analysis parameters and conditions

| Parameter | Condition |
|---|---|
| Column | (100 m × 0.25 m × 0.25 μm) |
| Injection volume | 1 μL |
| Split ratio | 1:25 |
| Head pressure | 44.695 psi |
| Flow | 1.2 mL $min^{-1}$ |
| Injector temperature | 250 °C |
| Oven profile | |
| Initial temperature | 100 °C |
| Hold time | 5 min |
| Ramp 1 | 7 °C $min^{-1}$ to 200 °C |
| Hold time | 2 min |
| Ramp 2 | 5 °C $min^{-1}$ to 230 °C |
| Hold time | 10 min |
| Detector | Flame ionization detector |
| Detector temperature | 250 °C |

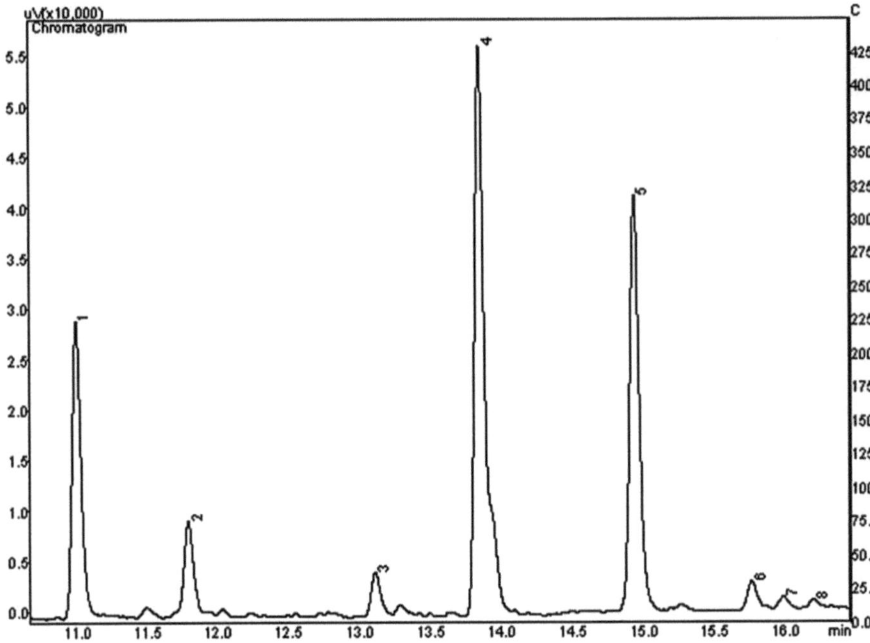


**Fig. 2.2** Fatty acid analysis by GC-FID. *1* palmitic acid (C16:0); *2* palmitoleic acid (C16:1); *3* stearic acid (C18:0); *4* oleic acid (C18:1*n9*); *5* linoleic acid (C18:2*n*6); *6* α-linolenic acid (C18:3*n*3); *7* eicosenoic acid (C20:1); *8* α-linolenic acid (C18:3*n*3)

Figure 2.2 shows a representative chromatogram of a fatty acids analyzed by gas chromatography–flame ionization detection (GC-FID), in order to characterize biodiesel's fatty acid profile.

#### 2.1.2.7 Acidic Index

One of the parameters of biodiesel quality is the content of free fatty acids, which is called acidic index. The acidic index is an important parameter to evaluate biodiesel corrosive potential. It can also be determined for a starting material (vegetable oil) as an indicative of the type of catalyst that should be used in its transesterification reaction (Xie et al. 2017). This value is expressed as amount of potassium hydroxide needed to neutralize a certain fraction of the oil or fat. ASTM and EN set a maximum of 0.50 mg of KOH $g^{-1}$ of biodiesel. Acidic index can be determined experimentally through a titration of biodiesel with aqueous KOH using phenolphthalein as indicator (Wang et al. 2008).

#### 2.1.2.8 Density

This property is related to the chain length and unsaturation index in the biodiesel fatty acids. Generally, vegetable oil has a greater density value when compared to transesterified oils. Density is a key aspect of biofuels, since such parameter is related to the quantity of combustible that enters the combustion chamber, which directly affects engine performance. The values of density are not set by the USA standards; however, it must be between 860 and 900 kg $m^{-3}$ at 15 °C according to the European Standards (Knothe and Razon 2017).

#### 2.1.2.9 Oxidative Stability

The unsaturated fatty acids of biodiesel are susceptible to oxidation upon exposure to oxygen, elevate temperature, light, metals, and peroxide. Oxidative stability affects biodiesel quality, since it is associated with storage conditions as well as antioxidant addition (Knothe 2007). In order to improve the oxidative stability of biodiesel, new stabilizing agents have been reported in the literature (Rockembach et al. 2015). Rancimat apparatus must be used to evaluate oxidative stability according to the EN and ASTM. For the EN, the biodiesel oxidative stability must be 8 h minimum. On the other hand, the ASTM set an oxidative stability of 3 h minimum.

#### 2.1.2.10 Glycerol Content

Glycerol (1,2,3-propanetriol) is the main by-product of biodiesel industry. The presence of small amounts of glycerol in a biofuel can cause several problems in an engine such as failure and coking (Bondioli and Bella 2005). According to EN and ASTM standards, free glycerol must be lower than 0.02% of the mass of the biodiesel.

As explained in this topic, the characteristics of biodiesel may vary according to several factors, such as fatty acid composition of the raw material, production process, conditions of storage and handling. Table 2.4 summarizes the chemical analyzes and its respective variation range accepted by standard agencies in Europe, USA, and Brazil (National Agency of Petroleum—ANP).

**Table 2.4** Specifications for B100

| Specification | ASTM | EN | ANP |
|---|---|---|---|
| Kinematic viscosity at 40 °C ($mm^2\ s^{-1}$) | 1.9–6 | 3.5–5 | 2.5–5.5 |
| Cetane number (min) | $\geq 47$ | $\geq 51$ | $\geq 42$ |
| Oxidation stability at 110 °C (h) | $\geq 3$ | $\geq 8$ | $\geq 6$ |
| Density at 15 °C ($kg\ m^{-3}$) | Not specified | 860–900 | 875–900 |
| Free glycerol (% mass) | $\leq 0.02$ | $\leq 0.02$ | $\leq 0.02$ |
| Monoacylglycerols (% mass) | $\leq 0.4$ | $\leq 0.7$ | Not specified |
| Total glycerol (% mass) | $\leq 0.24$ | $\leq 0.25$ | $\leq 0.38$ |
| Cold filter plugging point (°C) | Not specified | According to climate zone | Not specified |
| Acid value (mg KOH $g^{-1}$) | $\leq 0.5$ | $\leq 0.5$ | $\leq 0.8$ |
| Flash point (°C) | $\geq 93$ | $\geq 101$ | $\geq 100$ |
| Iodine value (g $I_2\ g^{-1}$) | Not specified | $\leq 120$ | Not specified |

## 2.2 Sources for Biodiesel Synthesis

### 2.2.1 *Algae*

Algae composes a very complex and diverse group of aquatic organisms, ranging from unicellular forms to multicellular, which can be up to 80 meters long. They are important components of many environments, which vary from tropical oceans to cold ecosystems, such as those from the Antarctic region. Eukaryotic algae are primarily divided according to its pigment components, and there are currently nine divisions (Vassilev and Vassileva 2016). The most important are Rhodophyceae (red algae), Chlorophyceae (green algae), Phaeophyceae (brown algae), Pyrrophyceae (dinoflagellates), Chrysophyceae (golden-brown algae), and Bacillariophyceae (diatoms) (Harwood and Guschina 2009).

These organisms are known to have several classes of lipids in its constitution, such as phytosterols, carotenoids, and fatty acids. Alongside, numerous other factors make algae biomass a very feasible source to produce biodiesel, because of its abundance, high rate of reproduction, and the fact that most of it is not used for human consumption (Nautiyal et al. 2014; Vassilev and Vassileva 2016). However, synthesis of biodiesel through this renewable source is still very expensive compared to other matrices, due to a costly cultivation, harvesting, separation, and the processing step. Moreover, the extraction process of its oil is not selective to fatty acids and, thus, extracts contaminants that may impact in the yield of the biodiesel synthesis, so as its purity and physicochemical properties (Viêgas et al. 2015).

In order to initiate the synthesis of biodiesel from algae biomass is necessary a previous lipid extraction step. In this procedure, organic solvents interact with the membranes of the organism in which fatty acids can be found. For an effective extraction, it is not only necessary to use a solvent that dissolves the lipids readily, but also can overcome the interactions from the lipids and the tissue matrix (Viêgas et al. 2015). Several extraction methods, which employ a multiplicity of organic solvents, have been proposed in the literature. Usually, they are based on the extraction potential of organic solvents or a mixture of these, followed by a phase separation and, finally, an isolation of the lipids. Most lipid extraction methods use a mixture of organic solvents with the endogenous water in the tissue as a ternary component of the system to extract lipids from the sample (Mandal and Mallick 2009).

In this work, microalgae *Scenedesmus obliquus* was evaluated for its potential as a raw material to the production of biodiesel. The fatty acid profile of the biodiesel from *S. obliquus* analyzed by GC-FID is shown in Table 2.5. The identification of the peaks from the chromatogram was made throughout a comparison to a standard comprised of a mixture of 37 fatty acid methyl esters (*Supelco TM*).

According to the results seen at Table 2.5, linoleic, oleic, and palmitic acids were the most abundant fatty acids found in the sample. However, other compounds, such as palmitoleic, stearic, and linoleic acids, were found in small quantities. The fatty acid composition in microalgae is due to endogenous and exogenous factors, such as genetic, temperature, and salinity that may induce the microalgae to produce certain types of fatty acids (Nautiyal et al. 2014). The proportion between unsaturated and saturated fatty acids is an important parameter to determine the quality of the biodiesel synthesized. A prevalence of unsaturated fatty acids may be easily oxidized and solidified at lower temperatures. In this case, biodiesel from *S. obliquus* proved to overcome this parameter (Guldhe et al. 2015).

Although not investigated in this work, several reports from the literature indicate that biodiesel from algae can easily overcome some specified parameters from standard agencies that control the quality of biodiesel. Usually, biodiesel from

**Table 2.5** Fatty acids (FAs) (% of area) found in biodiesel from algae

| Fatty acid | Quantity (%) |
|---|---|
| Palmitic acid (C16:0) | 18.3 |
| Stearic acid (C18:0) | 2.9 |
| Palmitoleic acid (C16:1) | 6.9 |
| Oleic acid (C18:1*n*9) | 39.8 |
| Eicosenoic acid (C20:1) | 1.9 |
| Linoleic acid (C18:2*n*6) | 26.5 |
| γ-Linolenic acid (C18:3*n*6) | 2.4 |
| α-Linolenic acid (C18:3*n*3) | 1.3 |
| ∑ Saturated FAs | 21.2 |
| ∑ Monounsaturated FAs | 48.6 |
| ∑ Polyunsaturated FAs | 30.2 |

**Table 2.6** Standards values and test results of biodiesel from algae *Scenedesmus obliquus*

| Parameter | Literature range | ASTM | EN | ANP | References |
|---|---|---|---|---|---|
| Density (kg $m^{-3}$) | 877 | 860–900 | 860–900 | 875–900 | Guldhe et al. (2015) |
| Saponification index (mg KOH $g^{-1}$) | 195.03 | Not specified | Not specified | Not specified | Wu and Miao (2014) |
| Iodine index (g $I_2$ 100 $g^{-1}$) | 71.62–98.68 | Not specified | $\leq 120$ | Not specified | Wu and Miao (2014), Guldhe et al. (2015) |
| Cetane number | 51.77–58.17 | $\geq 47$ | $\geq 51$ | Not specified | Wu and Miao (2014), Guldhe et al. (2015) |
| Acid index (mg KOH $g^{-1}$) | 0.42 | $\leq 0.8$ | $\leq 0.5$ | $\leq 0.5$ | Guldhe et al. (2015) |
| Oxidative stability (h) | 3.54 | $\geq 3$ | $\geq 6$ | $\geq 6$ | Guldhe et al. (2015) |

microalgae is capable of attending some specifications of biodiesel, such as viscosity and density, due to a considerable concentration of medium-sized chain monounsaturated and polyunsaturated fatty acids in its composition (Guldhe et al. 2015; Viêgas et al. 2015). The standards and the test's results of biodiesel from algae can be seen at Table 2.6.

According to Table 2.6, biodiesel from algae met several important parameters set by standard agencies in Europe, USA, and Brazil, such as viscosity, iodine index, and acid index. Therefore, according to data previously reported in the literature, microalgae *S. obliquus* proved to be a great source for the production of biodiesel, especially because of its fatty acid profile and the balance between unsaturated and saturated fatty acids.

## 2.2.2 *Castor Oil*

The castor oil plant (*Ricinus communis L.*) is a tree from hot weather environments, which makes India, China, and Brazil the biggest growers. Fifty percent of the castor seed total weight is a crude oil which has several applications in the pharmaceuticals, polymers, and lubricants industry. Regarding its potential, currently this plant is being used to produce biodiesel and, therefore, aiming to substitute edible vegetable oil biodiesel production (Dias et al. 2013; Azad et al. 2016).

According to the previous studies, the oil from the castor seed is mainly composed of ricinoleic acid. The massive presence of this compound gives this oil several unique features, such as high viscosity, increased polarity, stability, and a considerable hygroscopicity. These characteristics are due to the presence of an additional hydroxyl group in its carbon chain (Armendáriz et al. 2015). However, these seeds are poisonous and may cause health problems to individuals who deal with the plant and, thus, making it wide use difficult (Bateni et al. 2014). However, it is still expected that this second-generation biodiesel can overcome the currently dominating edible oil sources, due to the fact that these new fat alternatives also address social, economic, and environmental challenges beyond energy production (Azad et al. 2016). Figure 2.3 shows the chemical structure of ricinoleic acid.

Although very cheap and sustainable, castor oil is a very challenging matrix to produce biodiesel regarding its unique physicochemical properties discussed previously. Moderate solubility of ricinoleic acid in alcohols makes the transesterification process more practical to produce biodiesel. The literature indicates several ways to transesterify fatty acids from castor oil mainly using short-chain alcohols and alkaline catalysis. However, several factors may interfere with such process, for example, different catalysts, temperature, molar ratio between oil and alcohol, and reaction time (Armendáriz et al. 2015).

According to the literature, the kinematic viscosity of biodiesel from castor oil at 40 °C ranges between 18.3 and 60.9 $mm^2\ s^{-1}$. This value is higher than the EN and ASTM standards which indicate a viscosity at least smaller than 5 and 6 $mm^2\ s^{-1}$, respectively. Thus, this biodiesel may cause insufficient atomization in automobiles' engines leading to soot and deposits (Knothe and Razon 2017).

The main reason for this high value of kinematic viscosity from castor oil is related to the prevalence of ricinoleic acid in its fatty acid profile, since this molecule has a hydroxyl group which is soluble in the main by-product of the biodiesel synthesis. This fact makes purification of biodiesel more difficult and incomplete, which leads to a small presence of glycerol in the final product and, thus, an increased viscosity (Oliveira et al. 2013).

In this work, the biodiesel from castor oil was synthesized employing the basic catalysis and a yield close to 79% was obtained, a result that was influenced by the high density of the oil. Thus, increasing the time of the reaction and the stirring can be used to improve the synthesis of biodiesel. On the other hand, *p*-TSA as a catalyst had an 81% of yield. Therefore, this acid catalyst proved to be a good alternative to the conventional basic catalysis.

According to reports in the literature, biodiesel from castor oil failed in several parameters set by EN, ASTM, and ANP, as viscosity, density, and acid value. Based on that, this green combustible should not be used in its integral form. On the other hand, it can be employed blended to diesel or to other types of biodiesel. On the other hand, biodiesel from castor oil met several important parameters, such as iodine value, flash point, and cetane number. The standards and the test results of biodiesel from castor oil can be seen at Table 2.7.

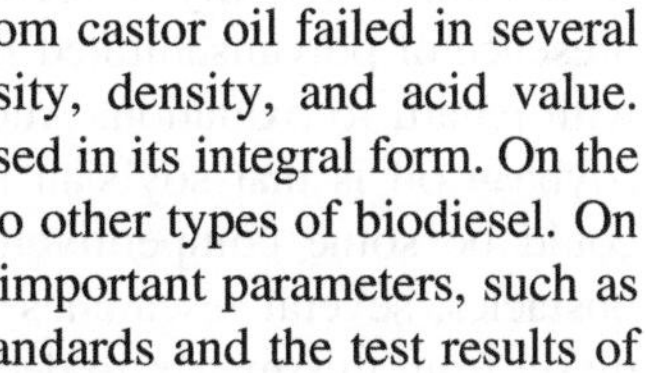


**Fig. 2.3** Chemical structure of ricinoleic acid

**Table 2.7** Standards values and test results of biodiesel from castor oil

| Parameter | Literature range | ASTM | EN | ANP | References |
|---|---|---|---|---|---|
| Viscosity ($mm^2\ s^{-1}$) | 18.3–60.9 | 1.9–6 | 3.5–5 | 3–6 | Armendáriz et al. (2015) |
| Saponification value (mg KOH $g^{-1}$) | 179.3 | Not specified | Not specified | Not specified | Sánchez et al. (2015) |
| Iodine value (g $I_2$ 100 $g^{-1}$) | 80.5 | No limit | $\leq 120$ | No limit | Sajjadi et al. (2016) |
| Acid value (mg KOH $g^{-1}$) | 0.70–1.19 | $\leq 0.5$ | $\leq 0.5$ | $\leq 0.5$ | Sánchez et al. (2015), Sajjadi et al. (2016) |
| Flash point (°C) | 147 | $\leq 93$ | $\leq 101$ | $\leq 100$ | Bateni et al. (2014) |
| Density (kg $m^{-3}$) | 936–950 | 860–900 | 860–900 | 875–900 | Dias et al. (2013), Armendáriz et al. (2015) |
| Cetane number | 42.3–60.4 | $\leq 47$ | $\leq 51$ | Not specified | Azad et al. (2016), Sajjadi et al. (2016) |

### 2.2.3 Soybean Oil

The soybean (*Glycine Max L.*) is a species that can be found in several countries with subtropical and tropical climates; thus, USA, Brazil, and Argentina are its largest production countries. This plant is widely known for its high amounts of oil and protein—fact that made it one of the most produced crops around the world. Since biodiesel synthesis is spiking every year, this species was one of the first to be used to produce biodiesel. For instance, in the USA alone, biodiesel jumped from 50,000 gallons in 1999 to 450 million of gallons in 2006 and keeps growing nowadays. Therefore, a considerable amount of oil was needed and soybean showed as one of the most promising (Chand et al. 2009).

According to previous studies, soybean oil is mainly composed by the polyunsaturated fatty acid known as linoleic acid (approximately 53% in weight) with a minor quantity of oleic, palmitic, and linolenic acids. However, the massive presence of polyunsaturated fatty acids makes biodiesel from this oil very unstable with regard to oxidation. Another limitation to the employment of biodiesel from soybean oil is that soybean is widely used for nutritional purposes, and so there could be some competitiveness for this feedstock. In order to overcome these obstacles, several researches are being made to introduce antioxidants in the biodiesel from this matrix to decrease oxidative instability (Rockembach et al. 2015).

Usually, biodiesel from soybean oil is made according to an alkaline catalysis and convention heating. Currently, there is a great interest to produce these green combustible employing alternative sources of energy like microwaves or ultrasound since they decrease the time of the reaction so as to increase its conversion. On the other hand, waste cooking oil has been extensively explored for the production of biodiesel gradually replacing virgin soybean oil. This is due to the fact the waste cooking oil is cheaper and is environmental-friendly (Boey et al. 2012).

Soybean oil biodiesel's kinematic viscosity, iodine index, and flash point were evaluated in order to characterize the obtained fuel. Results from this tests along with similar others found in the literature can be seen in Table 2.8. According to our experimental results and reported in the literature, biodiesel from soybean oil met several important parameters set by standard agencies in Europe, USA, and Brazil. Among these are viscosity, iodine value, acid value, flash point, total glycerol, and cetane number. However, this type of biodiesel failed in free glycerol and oxidative stability, due to its fatty acid profile. In other words, antioxidants should be added to biodiesel from soybean oil in order to increase its stability.

### *2.2.4 Waste Frying Oil*

Waste oils may include a variety of low-value materials such as used cooking or frying oils, and other waste materials. In most of hotels, restaurants, and in other food industries, the waste cooking oil is either simply discharged into the river or dumped into the land. Used or waste frying can be obtained from the restaurant industry, and they range from free up to 60% less expensive than neat vegetable oils, depending on the source and availability (Predojevic 2008).

The recent concerns with sustainability, environment, and raw material costs have made the use of waste frying oils attractive to the industry, especially if stimulated with economic benefits associated with the use of blends of biodiesel and fossil diesel (Zhang et al. 2003). Hydrolytic and oxidative reactions cause degradation on used vegetable oils, resulting in changes in the chemical and physical properties when compared to neat oils, such as increases in oil viscosity and acidity, color, odor, and aspect (Costa et al. 2000). Waste oils are normally characterized by relatively high free fatty acids, water contents, and solid materials that must be removed before biodiesel synthesis.

Waste cooking oil contains a high percentage of free fatty acid, making the alkaline reaction not suitable for biodiesel production from feedstock. In other words, because the base reacts with free fatty acid and forms soaps by saponification reaction, biodiesel conversion is reduced. In most of the cases, the waste oil must be filtered to remove the solid particles. The free fatty acids, however, have not been removed prior the biodiesel synthesis, since scientists have described the two-step transesterification reaction (Canakci and Van Gerpen 2001). In this reaction, the feedstock is treated usually with sulfuric acid to convert free fatty acids to the corresponding fatty acid methyl ester and followed by the transesterification reaction with a strong base as catalysts, usually sodium hydroxide or potassium hydroxide, to produce biodiesel (Naik et al. 2008). In summary, despite the additional cost of the two-step procedure, it readily accommodates high free fatty acid-containing low-cost feedstocks for the preparation of biodiesel (Canakci and Van Gerpen 2003).

**Table 2.8** Standard values and test results of biodiesel from soybean oil

| Parameters | Experimental value | Literature range | ASTM | EN | ANP | References |
|---|---|---|---|---|---|---|
| Viscosity ($mm^2 s^{-1}$) | 4.2 | 4.1–4.2 | 1.9–6 | 3.5–5 | 3–6 | Ramos et al. (2009), Sajjadi et al. (2016) |
| Iodine index (g $I_2$ 100 $g^{-1}$) | 110 | 118–128 | Not specified | $\leq 120$ | Not specified | Ramos et al. (2009), Sajjadi et al. (2016) |
| Acid index (mg KOH $g^{-1}$) | – | 0.14–0.18 | $\leq 0.5$ | $\leq 0.5$ | $\leq 0.5$ | Ramos et al. (2009), Sajjadi et al. (2016) |
| Flash point (°C) | 188 | 140–171 | $\geq 93$ | $\geq 101$ | $\geq 100$ | Ramos et al. (2009), Sajjadi et al. (2016) |
| Cetane number | – | 44–49 | $\geq 47$ | $\geq 51$ | Not specified | Ramos et al. (2009), Sajjadi et al. (2016) |
| Oxidative stability (h) | – | 1.3 | $\geq 3$ | $\geq 6$ | $\geq 6$ | Ramos et al. (2009) |
| Free glycerol (% mass) | – | 0.07 | $\leq 0.02$ | $\leq 0.02$ | $\leq 0.02$ | Ramos et al. (2009) |
| Total glycerol (% mass) | – | 0.00 | $\leq 0.24$ | $\leq 0.25$ | $\leq 0.35$ | Ramos et al. (2009) |

Based on that, biodiesel from waste frying oil was synthesized by a two-step process and analyzed for its quality parameters. The physicochemical properties evaluated on this work aim to comprehend the biodiesel quality and composition. The results are presented in Table 2.9. In summary, the biodiesel produced on this work showed similar values to the presented literature for all the properties evaluated.

In general, the parameters analyzed were all according to the literature reports and the regulatory agencies. Based on that, catalysis method investigated was confirmed to be effective for biodiesel synthesis.

#### 2.2.4.1 Chrysomya Albiceps Larvae

Organic wastes, such as animal waste, household and commercial waste, mainly restaurant waste, are generated daily in large quantities, which can cause pollution and induce major health risks if not treated correctly (Li et al. 2011). Restaurants are one of the most responsible for the waste generation. In Japanese restaurants, salmon, for example, is discarded practically whole since there is no used to produce the dishes.

**Table 2.9** Physicochemical parameters of biodiesel produced from waste frying oil

| Parameter | Synthesized biodiesel | Literature |
|---|---|---|
| Saponification index (mg KOH 100 $g^{-1}$ of sample) | 208.2 | 207 (Yaakob et al. 2013) |
| Acidic index (mg KOH $g^{-1}$ of sample) | 0.08 | 0.15–0.47 (Felizardo et al. 2006) |
| Iodine index (mg $I_2$ 100 $g^{-1}$ of sample) | 114 | 123–130 (Felizardo et al. 2006) |
| Viscosity ($mm^2$ $s^{-1}$) | 4.47 | 4.1–6.3 (Felizardo et al. 2006) |
| Flash point (°C) | 155 | 160–170 (Felizardo et al. 2006) |

This substrate is ideal for the development of several species in the form of scavengers insect. The production of biodiesel from flies larvae has been previously reported in the literature (Li et al. 2011). In this context, adult specimens of *C. albiceps* are attracted and may develop in this development medium. Lipids play an extremely important role in insects, in which they are constituents of cellular structures, act as hormones, and are important energy sources, where it is related to an insect flight activity. Due to their high efficiency, higher breeding rate, and shorter life cycle than plants, insects demonstrated the potential to substitute plants as raw material for biodiesel production (Li et al. 2012).

Based on the important relationship between lipids and insects, the objective of this study was to evaluate the lipid profile of *C. albiceps* fed with salmon head carcass from a restaurant to evaluate its use as biodiesel raw material.

To the development of the larvae, specimens of *C. albiceps* were collected in an urban setting, actively. After collection, the individuals were kept in plastic and transparent cages. They were fed diets based on sugar and protein, where they were kept under controlled temperature (±25 °C), relative humidity (70 ± 10%), and photoperiod of 12 h. The posture stimulus was performed by placing salmon carcass inside the cages for approximately 2 h. After that period, each meat with posture was removed from the cages and placed in a double organza closed system containing wet sawdust. As the larvae left the development medium, the pre-pupae were removed, washed with distilled water, dried on filter paper, and sacrificed at −4 °C.

Lipid extraction was performed following Soxhlet methodology. To this method, 10 g of *C. albiceps* in the pre-pupae stage were macerated and conditioned in the cartridge for extraction in the Soxhlet (hexane, 700 mL) under reflux (6 h). After the time, the solvent was evaporated under vacuum with 5% of yield. The biodiesel synthesis was performed according to a based catalyzed methodology described by Hartman and Lago (1973).

Biodiesel derived from *C. albiceps* was obtained in a moderate yield (50%). The Soxhlet extraction method was presented as an inexpensive method to obtain fatty

**Table 2.10** Fatty acid composition of *C. albiceps* pre-pupae biodiesel (% in area of total fatty acids)

| Fatty acid | Quantity (%) |
|---|---|
| Myristic acid (C14:0) | 1.69 |
| Pentadecanoic acid (C15:0) | 0.22 |
| Palmitic acid (C16:0) | 17.17 |
| Palmitoleic acid (C16:1) | 3.42 |
| Heptadecanoic acid (C17:0) | 0.16 |
| Stearic acid (C18:0) | 2.66 |
| Oleic acid (C18:1*n*9) | 38.27 |
| Linolelaidic acid (C18:2*n*6t) | 0.19 |
| Linoleic acid (C18:2*n*6) | 18.25 |
| $\gamma$-Linolenic acid (C18:3*n*6) | 0.27 |
| Linolenic acid (C18:3*n*3) | 6.63 |
| Arachidic acid (C20:0) | 0.24 |
| cis-11-Eicosenoic (C20:1) | 1.06 |
| cis-11,14-Eicosadienoic (C20:2) | 0.92 |
| cis-11,14,17-Eicosatrienoic (C20:3*n*3) | 0.40 |
| cis-8,11,14-Eicosatrienoic (C20:3*n*6) | 0.36 |
| Arachidonic acid (C20:4*n*6) | 0.51 |
| cis-5,8,11,14,17-Eicosapentaenoic (C20:5*n*3) | 2.66 |
| Behenic acid (C22:0) | 0.88 |
| Erucic acid (C22:1*n*9) | 0.22 |
| cis-4,7,10,13,16,19-Docosahexaenoic (C22:6*n*3) | 3.36 |
| Lignoceric acid (C24:0) | 0.23 |
| Nervonic acid (C24:1) | 0.24 |
| $\sum$ Saturated FAs | 23.25 |
| $\sum$ Monounsaturated FAs | 43.21 |
| $\sum$ Polyunsaturated FAs | 33.55 |

acids. Fatty acid profile of biodiesel from *C. albiceps* was analyzed by GC-FID, and the results are shown in Table 2.10. These data showed a balance of the fatty acids (saturated FAs 23.25%, monounsaturated FAs 43.21%, polyunsaturated FAs 33.55%), making a starting material for future research into biodiesel.

## 2.3 Conclusion

Different raw materials for biodiesel synthesis were presented. Different chemical routes for the preparation of biodiesel have also been shown. Certain materials such as soybean, castor and frying oils are commonly used as sources for biodiesel production and their physicochemical properties have been already determined.

However, there are alternatives that still deserve further research, such as algae and fly larvae. However, it is imperative to mention that biodiesel comprises an important product with a high demand in the world. This theme is an object of study in research, with low cost and the main advantage of being a renewable fuel.

## References


Alleman TL, Fouts L, Chupka G (2013) Quality parameters and chemical analysis for biodiesel produced in the United States in 2011. National Renewable Energy Laboratory (NREL), Colorado

Armendáriz J, Lapuerta M, Zavala F, García-Zambrano E, del Carmen Ojeda M (2015) Evaluation of eleven genotypes of castor oil plant (Ricinus communis L.) for the production of biodiesel. Ind Crops Prod 77:484–490

Azad AK, Rasul MG, Khan MMK, Sharma SC, Mofijur M, Bhuiya MMK (2016) Prospects, feedstocks and challenges of biodiesel production from beauty leaf oil and castor oil: a nonedible oil sources in Australia. Renew Sust Energ Rev 61:302–318

Bateni H, Karimi K, Zamani A, Benakashani F (2014) Castor plant for biodiesel, biogas, and ethanol production with a biorefinery processing perspective. Appl Energy 136:14–22

Bazooyar B, Ghorbani A, Shariati A (2011) Combustion performance and emissions of petrodiesel and biodiesels based on various vegetable oils in a semi industrial boiler. Fuel 90:3078–3092

Boey PL, Ganesan S, Maniam GP, Khairuddean M (2012) Catalysts derived from waste sources in the production of biodiesel using waste cooking oil. Catal Today 190:117–121

Bondioli P, Bella LD (2005) An alternative spectrophotometric method for the determination of free glycerol in biodiesel. Eur J Lipid Sci Technol 107:153–157

Boog JAF, Silveira ELC, Caland LB, Tubino M (2011) Determining the residual alcohol in biodiesel through its flash point. Fuel 90:905–907

Canakci M, Van Gerpen J (2001) Biodiesel production from oils and fats with high free fatty acids. Trans ASAE 44:1429–1436

Canakci M, Van Gerpen J (2003) Comparison of engine performance and emissions for petroleum diesel fuel, yellow grease biodiesel, and soybean oil biodiesel. Trans ASAE 46:937–944

Chand P, Reddy CV, Verkade JG, Wang T, Grewell D (2009) Thermogravimetric quantification of biodiesel produced via alkali catalyzed transesterification of soybean oil. Energy Fuels 23:989–992

Costa NPR, Rossi L, Zagonel G, Ramos L (2000) Produção de biocombustível alternativo ao óleo diesel através da transesterificação de óleo de soja usado em frituras. Quim Nova 23:531–537

Crizel MG, Lenz V, Ritter M, Pacheco B, Pereira CMP (2016) Viscosity analysis: a potential protocol to detect adulteration in biodiesel. Austin Environ Sci 1:1010

Dias JM, Araújo JM, Costa JF, Alvim-Ferraz MCM, Almeida MF (2013) Biodiesel production from raw castor oil. Energy 53:58–66

Ehsan M, Chowdhury MTH (2015) Production of biodiesel using alkaline based catalysts from waste cooking oil: a case study. Procedia Eng 105:638–645

Endo Y, Endo MT, Kimura K (2005) Rapid determination of iodine value and saponification value of fish oils by near-infrared spectroscopy. J Food Sci 70:C127–C131

Felizardo P, Correia MJN, Raposo I, Mendes JF, Berkmeier R, Bordado JM (2006) Production of biodiesel from waste frying oils. Waste Manag 26:487–494

Guldhe A, Singh B, Rawat I, Permaul K, Bux F (2015) Biocatalytic conversion of lipids from microalgae Scenedesmus obliquus to biodiesel using Pseudomonas fluorescens lipase. Fuel 147:117–124

Hartman L, Lago RCA (1973) Rapid preparation of fatty acids methyl esters. Lab Pract 22: 475–476

Harwood JL, Guschina IA (2009) The versatility of algae and their lipid metabolism. Biochimie 91:679–684
Hobuss CB, Rosales PF, Venzke D, Souza PO, Gobbi PC, Gouvea LP, Santos MAS, Pinto E, Lopes EJ, Pereira CMP (2011) Cultivation of algae in photobioreator and obtention of biodiesel. Braz J Pharmacogn 21:361–364
Ilkiliç C (2010) An analysis of exhaust emissions on a diesel engine operation by biodiesel. Energ Sources 33:298–330
Knothe G (2007) Some aspects of biodiesel oxidative stability. Fuel Process Technol 88:669–677
Knothe G, Razon LF (2017) Biodiesel fuels. Prog Energy Combust Sci 58:36–59
Knothe G, Steidley KR (2005) Kinematic viscosity of biodiesel fuel components and related compounds. Influence of compound structure and comparison to petrodiesel fuel components. Fuel 84:1059–1065
Li Q, Zheng L, Cai H, Garza E, Yu Z, Zhou S (2011) From organic waste to biodiesel: black soldier fly, Hermetia illucens, makes it feasible. Fuel 90:1545–1548
Li Z, Yang D, Huang M, Hu X, Shen J, Zhao Z, Chen J (2012) Chrysomya megacephala (Fabricius) larvae: a new biodiesel resource. Appl Energy 94:349–354
Lotero E, Liu Y, Lopez DE, Suwannakarn K, Bruce DA, Goodwin JG (2005) Synthesis of biodiesel via acid catalysis. Ind Eng Chem Res 44:5353–5363
Mandal S, Mallick N (2009) Microalga *Scenedesmus obliquus* as a potential source for biodiesel production. Appl Microbiol Biotechnol 84:281–291
Naik M, Meher LC, Naik SN, Das LM (2008) Production of biodiesel from high free fatty acid Karanja (Pongamia pinnata) oil. Biomass Bioenerg 32:354–357
Nautiyal P, Subramanian KA, Dastidar MG (2014) Production and characterization of biodiesel from algae. Fuel Process Technol 120:79–88
Oliveira DM, Ongaratto DP, Fontoura LAM, Naciuk FF, Santos VOB, Kunz JD, Marques MV, Souza AO, Pereira CMP, Samios D (2013) Obtenção de biodiesel por transesterificação em dois estágios e sua caracterização por cromatografia gasosa: óleos e gorduras em laboratório de química orgânica. Quim Nova 36:734–737
Predojevic ZJ (2008) The production of biodiesel from waste frying oils: a comparison of different purification steps. Fuel 87:3522–3528
Ramos MJ, Fernández CM, Casas A, Rodríguez L, Pérez Á (2009) Influence of fatty acid composition of raw materials on biodiesel properties. Bioresour Technol 100:261–268
Rashid U, Anwar F (2008) Production of biodiesel through optimized alkaline-catalyzed transesterification of rapeseed oil. Fuel 87:265–273
Rockembach CT, Pereira CMP, Dias D (2015) Evaluation of chalcones in the biodiesel oxidative process. Lambert Academic Publishing, Saarbrücken
Sajjadi B, Raman AAA, Arandiyan H (2016) A comprehensive review on properties of edible and non-edible vegetable oil-based biodiesel: composition, specifications and prediction models. Renew Sustainable Energy Rev 63:62–92
Sánchez N, Sánchez R, Encinar JM, González JF, Martínez G (2015) Complete analysis of castor oil methanolysis to obtain biodiesel. Fuel 147:95–99
Sandha GK, Swami VK (2009) Jojoba oil as an organic, shelf stable standard oil-phase base for cosmetic industry. Rasayan J Chem 2:300–306
Vassilev SV, Vassileva CG (2016) Composition, properties and challenges of algae biomass for biofuel application: an overview. Fuel 181:1–33
Viêgas CV, Hachemi I, Freitas SP, Mäki-Arvela P, Aho A, Hemming J, Smeds A, Heinmaa I, Fontes FB, da Silva Pereira DC, Kumar N (2015) A route to produce renewable diesel from algae: synthesis and characterization of biodiesel via in situ transesterification of Chlorella alga and its catalytic deoxygenation to renewable diesel. Fuel 155:144–154
Vieira BM, Elicker C, Nunes CFP, Bairros AV, Becker EM, Oliveira DM, Piva E, Fontoura LAM, Pereira CMP (2016) The synthesis and characterization of Butia capitata seed oil as a FAME feedstock. Fuel 184:533–535
Wang H, Tang H, Wilson J, Salley SO, Simon KY (2008) Total acid number determination of biodiesel and biodiesel blends. J Am Oil Chem Soc 85:1083–1086

Wu H, Miao X (2014) Biodiesel quality and biochemical changes of microalgae Chlorella pyrenoidosa and Scenedesmus obliquus in response to nitrate levels. Bioresour Technol 170:421–427

Xie WQ, Gong YX, Yu KX (2017) A rapid method for the quantitative analysis of total acid number in biodiesel based on headspace GC technique. Fuel 210:236–240

Yaakob Z, Mohammad M, Alherbawi M, Alam Z, Sopian K (2013) Overview of the production of biodiesel from waste cooking oil. Renew Sust Energ Rev 18:184–193

Zhang Y, Dube MA, McLean DD, Kates M (2003) Biodiesel production from waste cooking oil via two-step catalyzed process. Energ Convers Manage 48:184–188

# Chapter 3
# Alternative Multifunctional Additives for Biodiesel Stabilization: Perspectives for More Efficiency and More Cost-Effectiveness

**Ana Carolina Roveda and Magno Aparecido Gonçalves Trindade**

**Abstract** In this chapter, we address the current technologies to cover the alternative blends composed by multifunctional antioxidants—sometimes called secondary antioxidants—used to overcome the problems of degradation and provide more stability to the biodiesel which is derived from different raw materials. To this end, some valuable works with notable studies of conventional antioxidants, and sometimes synergetic binary/ternary blends and their applications, were briefly reviewed. However, the chapter attempts to cover only an overview of the recent advances in the field of multifunctional alternative additives, which provide a more efficient stabilization of the biodiesel. Also, its activities allow lowering the concentration of additives into biodiesel while maintaining their required specification. Furthermore, the chapter brings some aspects of the chemistry behind the multifunctional activities, focusing on the key benefits that afford in their multifunctional properties. Lastly, considering the vast scale in which the biofuel can be processed, we also show that the target additives improve the performance of conventional antioxidants more cost-effectively.

## 3.1 Introduction

As it is well-known, the search for alternative fuels was stimulated by the energy crisis of the 70's. As a consequence, the vegetable oils gained attraction as a potential biofuel for uses in diesel engines. In fact, the increased demand for energy in the industrialized world and the pollution problems caused by the widespread use of fossil fuels have also contributed to the need for new and renewable energy sources with less environmental impacts. Thus, biodiesel production has undergone

A. C. Roveda · M. A. G. Trindade (✉)
Faculty of Exact Sciences and Technology, Federal University of Grande Dourados, PO Box 364 Highway Dourados-Itahum, km 12, 79804-970 Dourados, MS, Brazil
e-mail: magnotr@gmail.com; magnotrindade@ufgd.edu.br

 
M. Trindade (ed.), *Increased Biodiesel Efficiency*, Green Energy and Technology,
https://doi.org/10.1007/978-3-319-73552-8_3

rapid and extensive technological reforms in industries and academies as the regulations were developed and this promoted the use of green fuels in the market and the increase of its commercialization.

Currently, one key search through biodiesel arises in the studies to improve its performance based on its physicochemical characteristics in order to maximize its stabilization. There are vast researches in this field, and many of them were previously reviewed. There is a drawback pointed for biodiesel production, though. Unlike fossil diesel, the methyl esters from vegetable oils, animal fats, and other alternative feedstock consisting mainly of triacylglycerols with its chemical nature make it more susceptible to oxidative processes and lower performance under cold conditions (Knothe 2007; Rasimoglu and Temur 2014; Knothe and Razon 2017). These issues are the most important in the evaluation of the biofuel quality because it involves the whole biodiesel logistics chain: from producer to final consumer. This is a fact that has led groups of researchers to solve these problems (Knothe 2007; Medeiros et al. 2014; Rasimoglu and Temur 2014; Knothe and Razon 2017). Cold flow and viscosity are two physical properties that are affected by poor feedstock quality. The problem associated with these properties is due to the fact that it can directly affect the biodiesel quality in its clean form and in blends with diesel. The main problems of operability are caused by the formation of solids and crystals that grow and agglomerate rapidly, creating problems in engine operation by choking fuel pipes and filters (Dwivedi and Sharma 2015; Knothe and Razon 2017). The cold flow properties are related to the amount of monoalkyl esters from the fatty acids present in the oils (Dwivedi and Sharma 2015). Viscosity is a measure of resistance to flow, and the biodiesel derived from feedstock with higher amounts of saturated chains in their composition tends to crystallize at lower temperatures. The main effect of high kinematic viscosity is to decrease the fluidity of the biofuel, which in turn, impairs the operationality of the fuel injection system. Thus, this parameter has been reported as one of the most harmful when using biodiesel blends in diesel engines. Eventually, this effect can be controlled by using oils with higher amounts of unsaturated fatty acids present in their composition. However, while the cold clogging temperature tends to decrease because of the lower content of saturated chains, the oxidative degradation of the biodiesel stability tends to increase due to the fact that the oxidative reactions easily take place in unsaturated fatty acids.

The low oxidative stability of biodiesel during the production and prolonged storage processes has been one of the drawbacks before its consolidation in the world energy matrix. Oxidative stability is reported as a parameter of biodiesel quality, and it is fundamental to evaluate the performance of automotive engines, especially, for increasing the useful life of parts, making the composition of fatty acids present in biodiesel the most important factor that influences their properties (Pullen and Saeed 2012). Thus, as there is an increase in the proportion of unsaturated fatty acids, biodiesel becomes more susceptible to oxidative processes (Cosgrove et al. 1987; Pullen and Saeed 2012). The process is directly related to the presence of allylic and bis-allylic carbons in the fatty acids that compose the biofuel. As the reaction is energetically favorable, the removal (from the radical) of a hydrogen atom from these groups leads to the formation of resonance-stabilized allylic and bis-allyl radicals (Medeiros et al. 2014). The formed radicals are susceptible to being attacked by atmospheric oxygen, converting into other radicals (peroxides). The peroxide radical reacts with the fatty acid present in the biodiesel forming the hydroperoxides, the primary products of oxidation (peroxides and hydroperoxides). In turn, its structures depend on the nature of the fatty acids which is present. As the enthalpy of this reaction stage is relatively low, this phase occurs more rapidly. The abstraction of the hydrogen from the fatty acid by the peroxide radical (ROO$^•$)—for the formation of the hydroperoxide radical (ROOH) and the radical (R$^•$)—preferably occurs in the carbon bonds where the energy of dissociation is low, such as the hydrogen of the bis-allyl position of the polyunsaturated fatty acid (Pokorny et al. 2001; da Silva et al. 2017). This stage results in an autocatalytic process, since the formed radicals act as propagators of the reaction. Fig. 3.1 shows how the biodiesel oxidation process happens. This process can be summarized in three main steps, commonly called as initiation, propagation, and termination (da Silva et al. 2017).

The presence of oxygen, water, and metals as well as the exposure to high temperatures and light incidence are factors that contribute to the reduction of oxidative stability of the biodiesel, which may trigger hydrolytic, photooxidative, and autooxidative reactions (Knothe 2007; Pullen and Saeed 2012; Fattah et al. 2014; Kivevele and Huan 2015; Knothe and Razon 2017). Transition metals have been reported as potent catalysts of biodiesel degradation processes even at trace-level amounts, leading to free radical formation that accelerates the oxidation of unsaturated fatty acids to produce primary and secondary oxidation products (Knothe 2007; Pullen and Saeed 2012). The presence of these contaminants in biodiesel can be derived from various sources, such as the storage and/or transport, during distillation and refining processes and also when blended with petrodiesel.

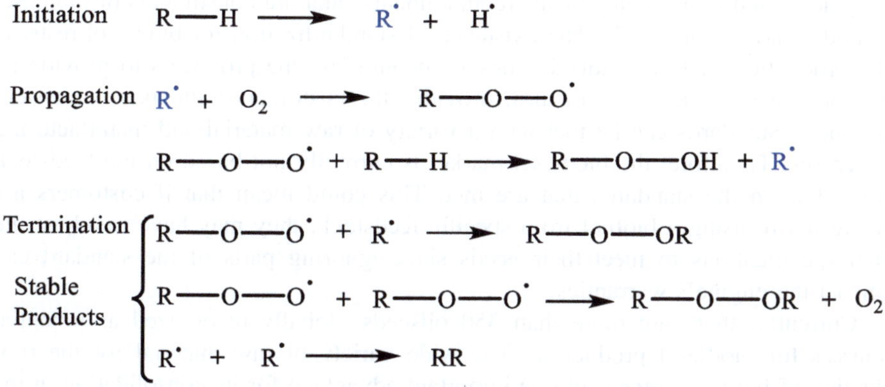


**Fig. 3.1** Schematic lipid oxidation reaction. RH: unsaturated fatty acid, R$^•$: free radical, ROO$^•$: peroxide radical, and ROOH: hydroperoxide

In this way, the knowledge of the chemical composition of biodiesel is fundamental to understand the causes of its instability.

Additives, such as antioxidants, have been introduced into biodiesel to block the formation of free radicals and ensure that over long periods of storage the biofuel meets with the quality demands from the market. Unfortunately, the addition of antioxidants into biodiesel also adds value to the final cost of production, making it less attractive than diesel. Thus, this chapter presents an overall overview of the action of natural and synthetic alternative multifunctional additives proposed to stabilize biodiesel (derived from several feedstocks), ensuring its improved quality in a more cost-effective way.

## 3.2 Biodiesel and Its Standardization

With the increasing interest in the use of methyl esters as biofuel, especially, since the end of the 90's, the establishment of quality standards is necessary to guarantee its quality. These include establishing the limits of contaminants that do not undesirably affect the quality of the emissions from the fuel burn, as well as the performance, the integrity of the engine, and the safety in the transport and handling.

Austria was the first country to define and approve the quality standards for biodiesel (e.g., biodiesel from rapeseed methyl esters). Subsequently, quality standards have been established in other countries. Currently, the American standard of quality, developed by American Society of Testing and Materials (ASTM) in the ASTM D6751, and the European Committee for Standardization (Comité Européen de Normalization—CEN) established in the European Union by the EN 14214 are recognized as the most well-known and are generally used as a reference or basis for other standards (Knothe 2005; Knothe and Razon 2017). These standards address a variety of fuel quality problems caused by monoalkyl esters from fatty acids and minor components (contaminants) that are constituents of biodiesel (Knothe and Razon 2017). The existence of standardization for biodiesel restricts the production of low-quality biodiesel, encouraging the producers to provide an appropriate biofuel, which in turn, provides the customers with more consistent products. Standards can be met with a variety of raw material and manufacturing processes. Therefore, the biodiesel market is currently not based on the feedstock used, but on the standards that are met. This could mean that if customers are interested in using a biofuel from specific feedstock, they may have to adjust the fuel specifications to meet their needs since ignoring parts of the standard can impact the engine's warranties.

Currently, there are more than 350 oilseeds globally recognized as potential sources for biodiesel production. The wide variety of raw material for the production of biodiesel represent one important advantage for its consolidation in the world energy matrix. The acquisition of feedstock is reported as being responsible for more than 75% of biodiesel production costs (Othman et al. 2017). Besides that, the cost of production can be significantly variable since it also depends on the production process, the place where it is stored and of course the amount of additives that are needed to maintain its stability (Abbaszaadeh et al. 2012). On the one hand, higher demand may stimulate oilseed production by increasing the scale of grain production and supply, which can result in lower costs. On the other hand, it can increase competition between food and fuel production, resulting in higher grain value, hence raising the cost of production. For example, rapeseed is the main source of biodiesel in Europe, while soybean oil is the most common source for biodiesel production in the USA, Argentina, and Brazil (Canakci and Sanli 2008; Mahmudul et al. 2017). To reduce costs, the use of animal and vegetable fatty wastes has been used for the production of biodiesel. However, even highly saturated esters can rapidly lose their stability if they contain a low amount of antioxidants.

Researchers have looked for alternative oil sources that present higher amounts of natural antioxidants, such as tocopherols and carotenoids, in their composition promoting greater stability for oxidation and oil quality (Grajzer et al. 2015; Smyk 2015; Lima et al. 2017). However, most of these natural antioxidants may be lost during the transesterification and refining processes of biodiesel (de Souza et al. 2017). In this way, the biodiesel quality may vary according to the molecular structures of its constituent esters or due to the presence of contaminants from the feedstock and the production process or formed during the biodiesel storage. The Rancimat method (raised by EN 14214) is the recognized standard method (method EN 14112) used for the analysis of the oxidative stability of biodiesel, whose minimum induction period of 8.0 h is required.

To meet the requirements of standards and to preserve the biodiesel quality for a long time, it is necessary to use effective additives. In this context, synthetic antioxidants have been widely used to preserve biodiesel, leading to an increase in its quality. However, it has been observed that the stability against oxidative degradation and the effectiveness of antioxidants are factors that are directly related to the final production costs. Thus, these issues have justified the growing interest in the research for newer and cheaper additives that can be applied to the biodiesel preservation, aiming to improve the effectiveness in a more cost-effective way.

## 3.3 Improving Biodiesel Stability with Conventional Antioxidants

As aforesaid, the quality of biodiesel is influenced by the oxidation of the fuel. As a consequence, properties such as kinematic viscosity, density, acid, and peroxide values are also impaired. The use of additives in order to retard oxidation processes and maintain the properties of biofuels can directly impact in the storage and distribution phases in the market. There are two types of additives: those that modify the physical characteristics, such as antiwear-lubricity, flavoring, coloring among others, as well as those that modify the chemical characteristics, such as antioxidants, anticorrosive, antirust, antifoams, detergent-dispersant (Rudnick 2013). In general, those whose proportion of addition to the product is given at extremely low levels in concentration are acceptable to the market.

Depending on their mode of action, antioxidants can be classified into groups such as free oxygen radicals and peroxides scavengers, metal-ion chelators (e.g., which catalyze oxidation of lipids), or light absorbers. The primary antioxidants act in two ways. The first one is interrupting the chain reaction by donating electrons or hydrogen to free radicals. These actions transform the antioxidant into thermodynamically stable compounds or react with free radicals, forming lipid–antioxidant complexes that react with other radicals (Adegoke et al. 1998; Pokorny et al. 2001; Pullen and Saeed 2012). The second one is the antioxidant acts in the initiation stage of the oxidative processes, through the complexation of metals, oxygen sequestration, decomposition of hydroperoxides to form non-radical species, deactivation of singlet oxygen or absorption of radiation (Pokorny et al. 2001; Pullen and Saeed 2012). In some cases, these agents can be used to increase the activity of primary antioxidants (Knothe 2007). The combination of two or more antioxidants is named synergism, in which the effect of the combination is greater than the sum of the individual antioxidant. Synergy may occur by the addition of two or more primary antioxidants or in a combination of primary and secondary antioxidants, being classified into homo- or hetero-synergism, respectively (Varatharajan and Pushparani 2017). The synergistic effects provided by mixtures of additives not only improve stability, but also allow a reduction in the amount of antioxidants needed. Thus, it directly influences in the final cost of biodiesel.

Antioxidants should be added as soon as the biodiesel is produced to maintain their properties and prevent the initiation of degradation processes. In order to have an effective stabilization in biodiesel, the additives must present some characteristics such as: absence of toxicity, low volatility, effectiveness in low concentrations, high thermal stability and light, high solubility in biodiesel, long duration in periods of storage, and low cost (Embuscado 2015). These additives can be classified as natural or synthetic antioxidants and are composed of phenolic groups, amines, or thiophenols. Much of the natural antioxidants have polyphenols, and they are similar to phenolic antioxidants. Due to potential sulfur emissions, thiophenols are not used to improve biodiesel stability. Furthermore, phenolic antioxidants are more efficient when compared to amine antioxidants (Varatharajan and Pushparani 2017). Among the natural ones, ascorbic acid, vitamin E, and β-carotene can be highlighted. Figure 3.2 shows five of the most important synthetic phenolic antioxidants that can be used, which are: *tert*-butylhydroquinone (TBHQ), butylhydroxytoluene (BHT), butylhydroxyanisole (BHA), propyl gallate (PG), and pyrogallol (PY) (Adegoke et al. 1998; Pokorny et al. 2001; Jakeria et al. 2014). The phenolic structure of these compounds allows the donation of a proton to a free radical, regenerating the acylglycerol molecule and disrupting the mechanism of free radical oxidation. In this way, phenolic derivatives are transformed into free radicals. However, these radicals can stabilize without promoting or propagating oxidation reactions.

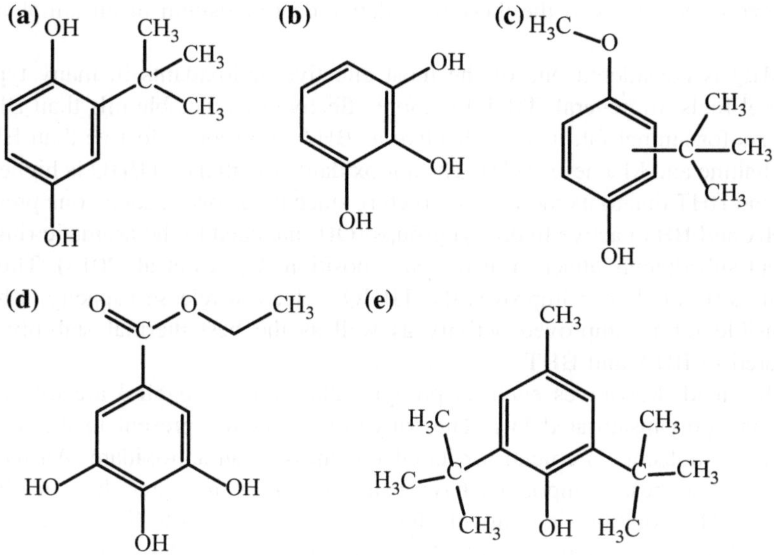


**Fig. 3.2** Chemical structures of **a** *tert*-butylhydroquinone (TBHQ), **b** pyrogallol (PY), **c** butylhydroxyanisole (BHA), **d** propyl gallate (PG), and **e** butylhydroxytoluene (BHT)

BHA is a more effective antioxidant in suppressing oxidation in animal fats than in vegetable oils. However, this antioxidant has been reported to have a suitable affectivity in the control the oxidation of short-chain fatty acids in some vegetable oils, such as those contained in coconut and palm oil (Maia et al. 2011). Like most phenolic antioxidants, the effectiveness of BHA is limited in unsaturated oils. The steric hindrance is responsible for a poor efficacy in vegetable oils. Nonetheless, this steric configuration helps protect the OH group, which is responsible for increasing thermal stability, especially, in animal fats (Min and Smouse 1985).

BHT has similar properties to BHA. Nevertheless, it is one of the most volatile antioxidants in phenolic groups. The rapid consumption resulting from high volatility and low thermal stability is the main disadvantages of BHT when added to biodiesel. The two *t*-butyl groups present on the BHT molecules affect their hydrogen donation ability caused by steric hindrance. Because it has less steric hindrance, the hydroxyl group of BHA is more favorable to the release of hydrogen than BHT. Thus, BHA exhibits more efficacy than BHT, not only in stabilizing biodiesel, but also in thermal stability, especially at 110 °C (Orives et al. 2014). Despite the low thermal stability and its steric hindrance, BHA presents better synergistic when added with gallates than BHT. However, both BHA and BHT are recognized to have a proper synergism with each other. While BHA acts as a sequester of peroxides radicals of unsaturated fatty acids, BHT acts synergistically (by the donation of electrons) to enhance the BHA effect (e.g., regeneration of BHA) (Omura 1995). The study highlighted by the authors shows that the structure of the antioxidant is fundamental to its activity (protective effect), especially in minimizing the effects of the thermo-oxidative decomposition of unsaturated fatty acids.

TBHQ is considered one of the most effective antioxidants in many types of biodiesel fuels. In general, TBHQ is more effective in vegetable oils than BHA or BHT. As for animal fat, it is as effective as BHA and more effective than BHT or PG (Chahine and Macneill 1974). The antioxidant potential of TBHQ is higher than BHA and BHT due to its molecular structure since it has two (against one presented by BHA and BHT) active hydroxyl groups (OH) attached to the aromatic ring with different substituents attached at the *para*-position (Orives et al. 2014). Thus, the lack of steric hindrance improves the TBHQ's electron release capacity, which is responsible for its improved activity as well as the best thermal stability when compared to BHA and BHT.

Gallic acid derivatives such as propyl gallate and pyrogallol are effective in stabilizing polyunsaturated fats. The trihydroxy structure present in the aromatic ring of the molecule is crucial for its effectiveness as an antioxidant. An inconvenience is that these compounds have better solubility in water than in oils and biodiesel. The volatility is relatively lower when compared to TBHQ, BHA, and BHT, and this property provides high thermal stability, besides being able to react with metallic ions. Gallates are effective in preventing the enzyme-induced oxidation of fatty acids (lipoxygenases) (Richard-Forget et al. 1995) and are reported to be more effective in stabilizing biodiesel. However, these compounds have a higher cost because the obtaining of the starting material is made from the plants (e.g., gallic acid), which in turn have been reported to be very costly (Fiege et al. 2000).

Researchers have found that the efficiency of a particular antioxidant depends on the feedstock (e.g., presence of natural antioxidants) and technology used in the production of biodiesel (e.g., pH control), which highlights the importance of conducting specific investigations for the esters produced by different matrices and reaction types (Mittelbach and Schober 2003). Mittelbach and Schober have studied the effect of the antioxidants PY, PG, TBHQ, BHA, and BHT on methyl esters produced with sunflower, waste frying oils, and animal tallow (distilled and not distilled). The antioxidant concentrations ranged from 100 to 1000 parts per million (ppm), and they used the Rancimat method to determine the efficiency of the additives. The authors found that PY, PG, BHA, and TBHQ significantly increased the stability of the studied biodiesel samples, whereas BHT was not very efficient even at higher concentrations than the other antioxidants. Jain and Sharma (2012) have also showed that the addition of PY led to a more pronounced effect followed by PG > TBHQ > BHT > BHA in increasing the stability of Jatropha biodiesel. Agarwal and Khurana (2013) demonstrated that for Karanja biodiesel the reported efficacy of various antioxidants tested was in the order of PY > PG > BHA > BHT > TBHQ.

Pullen and Saeed (2014) studied the efficiency of TBHQ in stabilizing biodiesel from four different types of biomass feedstocks (palm, olive, soybean, and Jatropha). It was verified that the biodiesel with high content of polyunsaturated fatty acids presented little improvement in the oxidative stability containing the same amount of antioxidant. The effect of temperature in the induction period

(IP) of biodiesel derived from animal fat, soybean oil, and cooking oil, using BHT as an antioxidant, was reported by Lapuerta et al. (2012) as a very important parameter to be evaluated. It was emphasized that at high temperatures (higher than 130 °C) the concentration of additive needed for meeting the required 8.0 h (specified in EN 14214) was significantly increased (higher than 25,000 ppm).

Pantoja and coworkers (Pantoja et al. 2013) evaluated the efficacy of BHA, PG, and TBHQ (at concentration level between 500 and 4000 ppm) in the stabilization of biodiesel produced from açaí (*Euterpe oleracea* Mart.), passion fruit, and linseed oil. It was showed that the difference in the level of unsaturation can significantly influence in the performance of the additive. The PG was considered the most effective antioxidant to stabilize the biodiesel from all studied feedstocks. For example, as far as the biodiesel obtained from açai is concerned, the PG was more effective even at lower concentration (500 ppm) than BHA and TBHQ, respectively. In a more significant effect, it was observed that their performance was enough to increase the induction period (IP) from 1.5 h (biodiesel without PG) to 21.3 h (biodiesel with PG). BHA was the second most effective antioxidant to stabilize the biodiesel from target feedstocks. However, to obtain a similar effect to PG, in some cases (e.g., biodiesel from passion fruit), it was necessary to double the amount of BHA.

Rawat et al. (2015) investigated the efficacy of individual and binary antioxidants to improve the oxidation stability of Jatropha and Karanja biodiesels. The antioxidant synergism was investigated using 500, 600, and 700 ppm for combination of the antioxidants, such as the mixtures of PY: PG, PY: TBHQ, and PY: BHA in different proportions. They observed that PY, PG, TBHQ, and BHA, when used individually, showed improvement in the induction period (IP) of biodiesel. However, a more significant improvement in the IP was observed using the binary mixture among these antioxidants, even when compared to the individual dosage at similar concentration. The highest concentrations of binary mixture produced negative synergy when the binary systems were composed by PY: PG and PY: TBHQ at proportion of 1:3 (weight/weight). Suitable synergism was observed. In the meantime, the binary mixture of PY: BHA resulted in antagonism in all concentrations and formulations studied. According to the authors, the efficiency of the binary system was found in the order of 1: 3/3: 1 > 1: 2/2: 1 > 1: 1 > 1: 9/9: 1.

As it was highlighted, the synthetic antioxidants are used as additive to improve the oxidation stability of biodiesel. However, depending on the physicochemical characteristics of biodiesel, large quantities of additives are needed to meet the specifications required by the standards. This leads to an increase in production costs. Synergistic combinations among these antioxidants can be used to improve the stability of biodiesel and to reduce the amount of required additives. Then, the ineffectiveness (e.g., as metal-ion chelators and light absorbers), low biodegradability, and toxicity of most of these antioxidants have been the focus of some publications. Thus, extensive research is needed to find more effective biodegradable alternative additives with low toxicity that work properly to preserve physicochemical characteristics, especially, in keeping the biodiesel quality more economic.

## 3.4 Improving Biodiesel Stability with Multifunctional Additives

Another drawback for improving biodiesel stability is that many of the commonly used antioxidants are very expensive. Then, perhaps the main challenges regarding the improvement of biodiesel stability using multipotent antioxidants and/or multifunctional additives are the cost-benefit ratio. Recent technological advances to obtain more efficient and more cost-effective additives have been observed and will certainly change this scenario. To overcome these challenges and understand how multifunctional additives can be effectively used to stabilize biodiesel, numerous synthetic and natural antioxidants have been proposed to be used individually or in combination (Fattah et al. 2014; Rawat et al. 2015; Varatharajan and Pushparani 2017).

For environmental and health reasons, there was a growing interest in the use of natural additives in both edible and biofuel derived from bio-based feedstocks. In general, vegetable oils present natural antioxidants such as tocopherols, tocotrienols, polyphenols, chlorophylls, ascorbates, lignins, and carotenoids that play an essential protective role in the oxidation of fatty acids (Varatharajan and Pushparani 2017). On the one hand, biodiesel produced from unrefined vegetable oils may contain more natural antioxidants and it can be more stable when compared to the biodiesel produced from other raw materials (Knothe 2007). On the other hand, the concern involving biofuel derived from target vegetable oils is due to the fact that many naturally occurring antioxidants may be destroyed or partially removed during the biodiesel production process. The loss of natural antioxidant also happens because most of the extracted vegetable oils (e.g., oilseed radish) need to be subjected to a degumming process to eliminate gums, phospholipids, among other substances that have direct impact on the quality and the yield of biodiesel (Diana da Silva Araújo et al. 2014; de Souza et al. 2017). To solve this shortcoming, it is still essential to use synthetic antioxidants within biodiesel to suppress its oxidation during storage processes.

### *3.4.1 Natural Multifunctional Additives*

Most phenolic compounds derived from plant have antioxidant properties. These include tocopherols, carotenoids, lycopene, gallic acid, caffeic acid, vanillin, synapic acid, ferulic acid, protocatechuic acid, *para*-coumaric acid, eugenol, vanillic acid, and cinnamic acid. Thus, most of these compounds can be commercially produced. Several studies have been done with plant extracts in search of compounds that present antioxidant, antitumor, anti-inflammatory, among other properties. Kreivaitis et al. (2013) tested the efficiency of sage and thyme extracts in preventing the oxidation of rapeseed oil during 70 days of storage. The addition of antioxidants prolonged the induction period of colza oils twice, with sage extract being more efficient than thyme extract. Extracts from sage, rosemary, clove, allspice, thyme, cinnamon, oregano, marjoram, eucalyptus, artichoke, and saffron, for example, are also reported as effective antioxidants in food products (Embuscado 2015). However, apart from some extract containing tocopherols, few studies have been conducted aiming the uses of these natural compounds as antioxidants in biodiesel.

Moser (2012) studied the efficacy of gossypol and γ-tocopherol in the inhibition of the oxidation of biodiesel produced from soybean oil, residual cooking oil, and technical grade methyl oleate. Unexpectedly, it was observed that natural antioxidants, even at the amount of 250 and 500 ppm, were less effective when compared with BHT. Moser (2008) also investigated the effectiveness of myricetin oil extracts, which is obtained from the *Moringa oleifera* seed, in soybean oil biodiesel and reported that myricetin has a better antioxidant activity than α-tocopherol. Most studies report that tocopherols have a limited antioxidant activity in biodiesel when compared to synthetic antioxidants. This limitation arises from the fact that tocopherols can be partially consumed before their action as antioxidant, because they are easily oxidized by air and also by the lipoxygenase enzyme present in most feedstocks (Håkansson and Jägerstad 1990).

Damasceno et al. (2013) tested the antioxidant activity of caffeic acid (CFA), ferulic acid (FA), and TBHQ in biodiesel to evaluate the effect on stability during the storage period. It was demonstrated that the efficiency of target antioxidants was in the order of: CFA > FA > TBHQ, in which the CFA at 1000 ppm was able to stabilize biodiesel (meeting with EN 14214 specifications) even after three months. The authors reported that in the hydrogen donation process by CFA its free radicals generated a dimer that has excellent antioxidant properties to provide additional protection. Nivetha and Roy (2013) evaluated the efficacy of natural and synthetic antioxidants such as BHT, 2,6-di-tert-butylphenol (DTBP), quercetin (QC), gallic acid (GA), methanol extracts of green tea (GT-M) and pomegranate hull (PH-M) and green tea chloroform extract (GT-C) on the biodiesel stability of *Hevea brasiliensis*. In general, the authors observed that the performance of synthetic antioxidants was better compared to their natural counterparts. The order of antioxidant activities in concentrations above 1500 ppm was: DTBP > BHT > GA > GT-C > GT-M > QC > PH-M, in which the DTBP was the most effective as a protection factor.

Curcumin has been reported as a potential antioxidant for biodiesel in replacing synthetic antioxidants. The promising activity was supported by its capacity to increase biodiesel stability by up to 83%. This protective activity is significantly better than BHT and β-carotene (De Sousa et al. 2014; Serqueira et al. 2015). In general, synthetic antioxidants are still more effective than natural antioxidants in improving oxidation stability, since lower amounts of additives are required to meet specifications when compared to natural antioxidants. To overcome these issues and other shortcomings, several studies have reported the use of binary and ternary combinations of various natural and synthetic antioxidants. In this sense, Niki and coworkers (Niki et al. 1984) evaluated the synergistic effect between α-tocopherol and ascorbic acid when added to methyl linoleate. The combination showed positive synergy, in which ascorbic acid was reported as an effective hydrogen donor to regenerate α-tocopherol. In this way, the mixture was better able to improve the biodiesel preservation than the individual additives.

Hraš et al. (2000) evaluated the synergistic effect of rosemary extract in combination with citric acid, ascorbylpalmitate, and α-tocopherol in sunflower oil. All combinations showed synergistic effects, and the mixture containing rosemary extract with ascorbylpalmitate was the most effective for the protection of sunflower oil. Later, Medeiros et al. (2014) in turn showed that the mixture containing the combination of ethanol extract (from rosemary) and TBHQ has positive synergism in stabilizing biodiesel produced with cottonseed oil. In addition, Deyab (2016) has shown that ethanol extracts from rosemary leaves reduce the corrosion rate of aluminum in biodiesel. Serrano et al. (2013) reported that citric acid-washed biodiesel showed better stability with all types of antioxidants. Spacino et al. (2016) have also shown a significant synergistic activity with the addition of alcoholic extracts of rosemary and oregano at a ratio of 1:1 in methyl ester of soybean.

Currently, natural occurrence antioxidants have not achieved remarkable commercial success mainly due to their higher acquisition cost and greater quantity needed to meet biodiesel quality parameters. Therefore, another important point to be considered in this issue is regarding the composition of biodiesel that can be significantly altered by the presence of high amount of additives. However, with the reported inefficiency of the primary antioxidants when used individually in preserving the biodiesel against degradation from exposure to metals and UV radiation, the search for natural multifunctional antioxidants derived from bio-based feedstocks is still on demand. In addition to poor ability to complex with metals, most synthetic antioxidants (e.g., such as BHA and BHT) can also volatilize or decompose at temperatures around 110 °C (temperature used in the Rancimat® test), leading to a wrong result of the oxidative stability test (Lamba et al. 2013).

In addition to the aforementioned, it is also reported that the conventional synthetic antioxidants, even when used in high concentrations, have not been effective in providing satisfactory results against degradation of biodiesel in the presence of heat or contamination by metals (Sarin et al. 2010a; Fattah et al. 2014; Sulistyo et al. 2015). The presence of metal ions, even at trace levels, may act as potent catalysts to accelerate the oxidation and polymerization reactions of unsaturated fatty acids. The reaction takes place because transition metals can act as a Lewis acid to catalyze the polymerization of polyunsaturated hydrocarbon molecules present in most feedstock, which are thus present in biodiesel (Banga and Varshney 2010). The occurrence of metal ions in biodiesel may not be fully avoided. For example, iron and copper may be present in biodiesel from the raw material and/or in the manufacturing process (e.g., if corrosion occurs in the storage tanks). Similarly, the alkali metals (responsible for the formation of sediments that cause injector failures) can originate from raw material (depending on the mode cultivation) and during the transesterification process. The consequences of the presence of alkali metals in biodiesel can also be very dangerous for the engines. For example, alkali metals accelerate the saponification process and contribute to the formation of insoluble or water mist in diesel blends (Banga and Varshney

2010). In this context, efforts should be placed in search for multipotent and/or multifunctional additives that work properly in overcoming these shortcomings and in being as economic as possible.

### *3.4.2 Alternative Multifunctional Additives*

Multipotent and/or multifunctional additives are defined as compounds containing groups with characteristics of acting further than the proton donation (which is raised by conventional antioxidants) to the radicals generated by the oxidation process. In recent years, several studies have been devoted to finding alternative additives in an attempt to combine radical-scavenging (or radical-generation-preventing) activity and other properties such as pro-oxidative metal inhibitors, singlet oxygen deactivation, oxidative enzyme inhibitors, and UV absorbers (Sarin et al. 2010b; Comin et al. 2017). Knowing this limitation, Sarin and coworkers (Sarin et al. 2010b) proposed the combination of BHT and N,N′-disalicylidene-1,2-diaminopropane to be added to the Jatropha biodiesel as suitable strategies to combine multiple functions for radical-scavenging (BHT) and metal deactivator properties (N,N′-disalicylidene-1,2-diaminopropane). In this study, the authors address the ability of metal deactivators in chelate transition metal ions to inhibit their catalytic effect on the oxidation of biodiesel.

Among the researches involving some naturally occurring antioxidants, most of the studies on the preservation of biodiesel are related to the synergistic effect exploring the properties of primary synthetic antioxidants (phenolic type) and secondary antioxidants (including some natural extracts). Researchers have shown that conventional antioxidants (primary antioxidants) can be combined with metal deactivating compounds (secondary antioxidants) to stabilize biodiesel contaminated with metals (Sarin et al. 2009, 2010a; Roveda et al. 2016; Comin et al. 2017). In general, these secondary antioxidants are added to regenerate the primary antioxidants during its action, leading to an increase in their action time. This issue has great cost-effectiveness importance, since it allows a decrease in the amount of primary antioxidants (that are usually expensive), while maintaining their activity during long-term action. Ethylenediaminetetraacetic acid (EDTA), diethylenetriaminepentaacetic acid (DTPA), sodium tripolyphosphate, N,N′-disalicylidene-1,2-diaminopropane, and citric acid are well-known classic chelating agents and used to solve the aforesaid limitations of the primary antioxidants (Varatharajan and Pushparani 2017). Binary and/or ternary mixtures containing conventional antioxidants and chelating agent can potentially be more suitable to provide a synergistic effect, leading to a higher effectiveness as pro-oxidative metal inhibitors that are present in biodiesel.

Another way to use binary and/or ternary mixtures and have satisfactory synergistic effects is regarding the combination of conventional antioxidants and alternative additives, the so-called solvent dyes or markers dyes (Trindade et al. 2007, 2010, 2011). The argument behind the use of these markers in biodiesel arose from the fact that in the last three decades dyes were used as fuel additives, especially in petroleum products, with the purpose of marking to differentiate their different types and quality (Trindade and Zanoni 2009). With the increasing quality control of petroleum and its derivatives, especially in the content of sulfur compounds (Trindade et al. 2014) and the possible adulterations with solvents (Romanini et al. 2009; Trindade et al. 2012), the guarantee of these products was associated with the use of markers dyes (Trindade et al. 2011). For environmental reasons, these specifications were even more important for diesel because it is one of the most critical petroleum derivatives. The marking practice makes it possible to monitor the widespread use of low-quality fuels (high sulfur content) in on-road vehicles as well as to control low taxation products—subsidized for exclusive use in off-road vehicles, heating oil among other purposes (Trindade et al. 2011). In view of that, researchers (Roveda et al. 2016, 2017; Comin et al. 2017; de Souza et al. 2017) proposed the use of some markers dyes (Picture 3) in the field of biodiesel stabilization.

The structures in Picture 3 show that the markers dyes have favorable structures to be used as biofuel additives due to the presence of groups of substituents with their hydrogen donation capability. Moreover, the presence of chromophores groups (color-bearing group) and also auxochromes groups (color helpers: substituent groups with electron donating capacity), makes it potentially attractive as biodiesel additives due to their very important characteristic as light absorbers. These groups may not only help in the hypsochromic shift but also in the ability to form metal-ion chelators. The chelators are generally water soluble, but some (e.g., citric acid) will exhibit solubility in lipids. For this reason, they are recognized as effective metal complexant in oleaginous samples (Decker et al. 2010). Thus, considering the fact that markers dyes, as shown in Fig. 3.3, are soluble in petroleum products, this characteristic can be explored to inactivate the metal ions in the biodiesel from the formation of soluble metal complexes. The formation of soluble metallic complexes with transition metal ions eventually present in biodiesel or in the fossil fuel plays an important role in the challenges to avoid catalytic biodiesel degradation caused by the contact with these metals (Comin et al. 2017).

Following the positive aspects aforesaid, Roveda and coworkers (Roveda et al. 2016) evaluated the thermodegradation of soybean biodiesel stabilized with the combination containing synthetic antioxidants (BHT and PG) and an alternative additive based on the anthraquinone dye (1,4-dihydroxyanthraquinone, DHQ). The authors showed that the combinations of antioxidants and DHQ were more effective in stabilizing the biodiesel than the individual addition at similar concentration. The same group (Roveda et al. 2017) also studied the efficacy of the combination containing the antioxidant TBHQ and the anthraquinone compound, named 1,4-bis (ethylamino)-9,10-anthraquinone (BEQ). All the proposed combinations between the additives presented positive synergism, allowing reductions in the amount of the primary antioxidant (TBHQ) in the mixture, leading to a lower cost than its individual addition.

de Souza et al. (2017) evaluated the effect of a binary combination, having TBHQ and a dye, the so-called solvent blue 35 (SB-35), on the thermal stability of

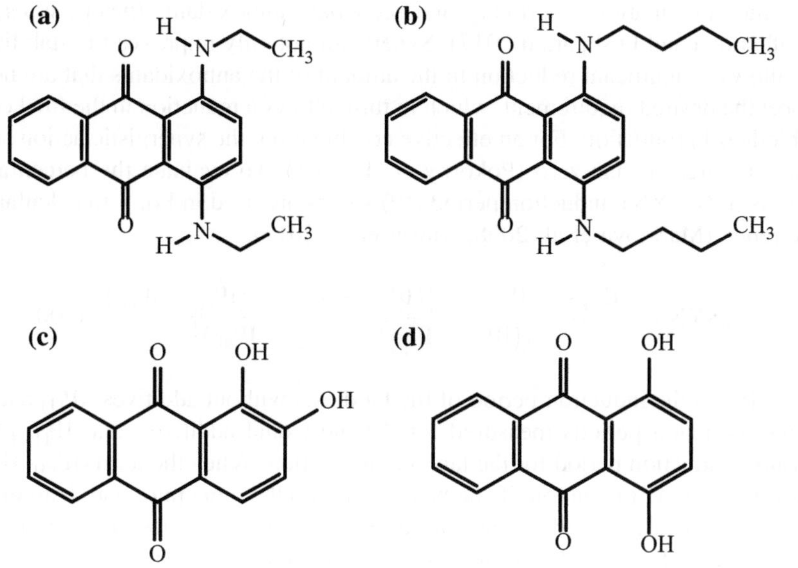


**Fig. 3.3** Chemical structures of **a** 1,4-bis (ethylamino)-9,10-anthraquinone (BEQ), **b** CI solvent blue 35 (SB-35), **c** alizarin (ALZ), and **d** quinizarin (QNZ)

oilseed radish (OSR) and the biodiesel from the extracted oilseed radish (OSRB). The study showed the importance of using the alternative combination on the field of biofuel stabilization, for which OSR and OSRB were more effectively stabilized when using the proposed combination. The mixture (TBHQ + SB-35) had strong antioxidant activity even at very low concentrations when compared to the addition (at similar concentration) of individual compounds. Comin et al. (2017), in turn, analyzed the effect of transition metal ions on the stability of soybean biodiesel (SB) in the presence and in the absence of binary and ternary mixtures. The combinations were composed by alizarin (ALZ) as the alternative additive, TBHQ and citric acid (CA) as the conventional antioxidants. They concluded that even at very low concentration, the metals promptly degrade the SB in the absence of additives. A similar effect was observed for the SB blended with the additives added individually. However, even at low concentrations, the proposed combination had a positive impact on SB stability compared to the effects of individual additives. These authors showed that this issue has great impact on the effectiveness of the cost because the amount of primary antioxidant TBHQ (expensive) was reduced without any negative impact on its activity during a long-term action.

Among the above researches, a way to calculate the effectiveness of the binary or ternary mixture is the use of a simple calculus to find the synergistic effect. As mentioned earlier, synergism is the combined use of two or more additives to produce a greater effect than the sum of any individual additive. In general, synergism can be achieved by the addition of primary antioxidants (homo-synergism) or by the combination of primary and secondary antioxidants (hetero-synergism) (Varatharajan and Pushparani 2017). Synergism not only improves the stability but also allows a significant reduction in the amount of the antioxidants that are needed to meet the desired requirement, which in turn, allows a reduction in the final cost of the biodiesel production. For an effective combination, the synergistic action should always be greater than zero (Pokorny et al. 2001). To evaluate the percentage of synergism (% SYN), induction period (IP) values are used in Eq. 1 to calculate this occurrence (Marinova et al. 2008; Rawat et al. 2015).

$$\%\mathrm{SYN} = \frac{\left(\mathrm{IP}_{[1,2]} - \mathrm{IP}_{[0]}\right) - \left[\left(\mathrm{IP}_{[1]} - \mathrm{IP}_{[0]}\right) + \left(\mathrm{IP}_{[2]} - \mathrm{IP}_{[0]}\right)\right]}{\left[\left(\mathrm{IP}_{[1]} - \mathrm{IP}_{[0]}\right) + \left(\mathrm{IP}_{[2]} - \mathrm{IP}_{[0]}\right)\right]} x100 \qquad (1)$$

where $IP_{[0]}$ is the induction period of the biodiesel without additives, $IP_{[1]}$ and $IP_{[2]}$ are the induction periods measured for the individual additives, and $IP_{[1,2]}$ is the measured induction period for the target combination. When the term ($IP_{[3]}$–$IP_{[0]}$) is used for ternary combination, it can be entered into the numerator and denominator, respectively. Thus, the authors have used this equation to evaluate the occurrence of synergism and they found that, in general, when values are positive (e.g., great than 20%), the combination has great ability to stabilize biodiesel, maintaining some quality parameters (oxidative stability, density, peroxide, and acid values) lower than the specified values.

## 3.5 The Chemistry Behind the Multifunctional Antioxidant Activity

The hydrogen-donating ability is one of the main factors (in antioxidants activity) to be considered for the screening of antioxidants (Varatharajan and Pushparani 2017). Primarily, the hydrogen bonding reaction takes place as a function of bond dissociation enthalpy (BDE) in which the ability of this action can predict if the additive is able to effectively neutralize free radicals. The bonding energy of the –OH and –NH groups of an antioxidant should be considerably lower than the values BDE of peroxides and free radicals of hydroperoxide (Varatharajan and Pushparani 2017). In some cases, other factors may also play an important role in determining the antioxidant effectiveness, as the presence of bulky groups near the −OH and −NH groups generates steric hindrance (Barclay and Vinqvist 2003). Likewise, the mobility and solubility to migrate freely in the medium are very important. Low molecular weight antioxidants are not recommended, since they are volatile in nature and the loss by evaporation (at moderated temperature) becomes inevitable.

Based on the previous work that covers the chemistry behind the antioxidant activities and/or synergetic effects (Choe and Min 2009; Pisoschi and Pop 2015; Yehye et al. 2015), the primary antioxidants such as TBHQ, BHA, BHT, and PG are mainly responsible for blocking the formation of hydroxyl free radical or superoxide radical anion via hydrogen donation. Secondary antioxidants such as citric and ascorbic acid, in turn, are responsible for promoting the decomposition of hydroperoxides, inhibitors of pro-oxidative metals, deactivation of singlet oxygen, oxidative enzyme inhibitors, UV radiation absorbers, among others. When combined, the occurrence of synergistic interaction has been noticed and for this reason, the antioxidants can regenerate each other.

According to the study developed by Comin et al. (2017), the insight into how a marker dye with a very low concentration is able to bring about the synergistic effect claimed with multifunctional additives was put forward. The authors have proposed a mechanism to explain how a diaminoanthraquinone derivative is able to enhance the effectiveness of a hydroquinone in stabilizing the substrate toward oxidation. The proposed mechanism (Fig. 3.4) shows that the regeneration may happen through a third additive, citric acid (CA). In this pathway, the CA becomes a hydrogen donor (i.e., pKa1 = 3.0, pKa2 = 4.7, pKa3 = 5.4) and, then, while the donor is consumed, the TBHQ propagates its oxidation inhibition (Fig. 3.4a). In the next step, the hydrogen donor CA can be regenerated by alizarin (ALZ, pKa = 7.0), in which, CA becomes able to donate one more hydrogen to the radical TBHQ. This mechanism shows that the hydrogen donation is favored thanks to the stability of ALZ radical by resonance (Fig. 3.4b). In the third step (second cycle of CA regeneration), the stabilization of CA happens because the radical from ALZ donates its second hydrogen and becomes a diketone and it can easily chelate with the metal ions (Fig. 3.4c).

In that mechanism, the synergistic effect observed with both secondary antioxidants (ALZ and CA) occurs because ALZ is able to regenerate the CA, which in turn is capable of continuing the regeneration (by hydrogen donation) of the primary antioxidant TBHQ. Thus, ALZ is claimed here as the multifunctional additive, since its function allows the donation of its hydrogen and consequently the formed radical is able to chelate the pro-oxidative metals. The process helps to maintain the activity of TBHQ in inhibiting the formation of fatty acid radicals, even at a low level of concentration and the presence of pro-oxidative metal.

## 3.6 Cost Estimation for Using Multifunctional Antioxidants

Generally, the most important influences on the cost of biodiesel production are highlighted as: (i) capital and operating costs of the plant (e.g., the buildings, the processing plant, services, catalyst as well as the feedstock and product storage), (ii) feedstocks used in the production process (e.g., tallow, vegetable and/or waste oil, and alcohol, typically ethanol or methanol), (iii) the glycerol byproduct (e.g., which can provide a secondary revenue stream), and (iv) the yields as well as the quality of the biodiesel (which may include stabilization process) and glycerol (Haas et al. 2006; Liu 2015). Although the price of stabilization is not addressed

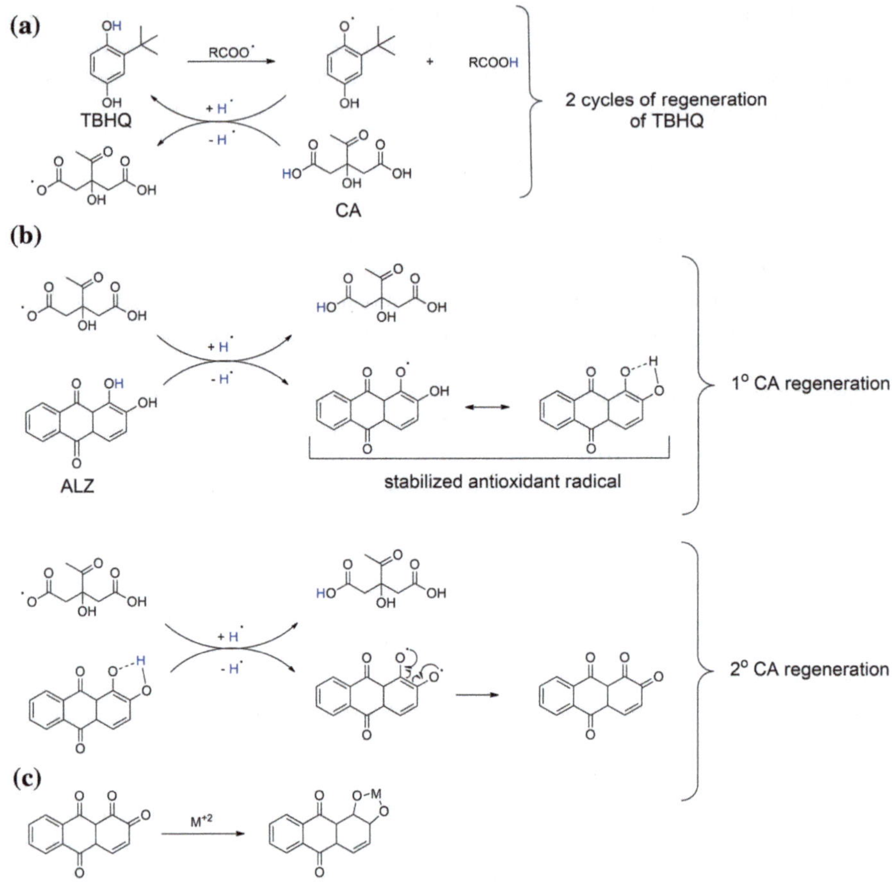


**Fig. 3.4** General schematic mechanisms for the synergistic interaction between the additives TBHQ, CA, and ALZ. **a** Cycle of regeneration of TBHQ, **b** cycle of regeneration of CA, and **c** chelation of diketone from ALZ with metals from metallic additives. Reproduced from Ref. (Comin et al. 2017) with permission from Elsevier. Copyright 2017, Elsevier Publisher

(by the authorities concerned) as one of the main components on the cost of biodiesel production, this issue can provide a basis on which the final cost of biodiesel production must be compared to.

Here, the cost of stabilization process was estimated at the time of writing this chapter and the basis for evaluating the cost of additives per ton of soybean biodiesel was the chemical values obtained from Sigma-Aldrich Company, USA. It can be seen from Table 3.1 that when each additive is compared to its combinations, cost varies significantly. It is worth noticing that, even after the addition of three compounds, the combination is still considerably less expensive than the standard antioxidants (e.g., TBHQ and PG). The large cost difference happens because in order to get target standards additives to have similar effects, they need

**Table 3.1** Cost of the additives per ton of soybean biodiesel samples doped with the combinations as well as with each additive individually

| Additives and concentrations | Cost per ton (USD) | Cost reduction (%) |
|---|---|---|
| TBHQ (500 mg $kg^{-1}$) | $187.50 | – |
| PG (500 mg $kg^{-1}$) | $223.00 | – |
| BHT (500 mg $kg^{-1}$) | $74.40 | – |
| TBHQ (100 mg $kg^{-1}$) + ALZ (50 mg $kg^{-1}$) | $43.50 | 77 |
| TBHQ (100 mg $kg^{-1}$) + ALZ (50 mg $kg^{-1}$) + CA (10 mg $kg^{-1}$) | $43.60 | 77 |
| PG (50 mg $kg^{-1}$) + DHQ (10 mg $kg^{-1}$) | $25.50 | 89 |
| PG (100 mg $kg^{-1}$) + DHQ (50 mg $kg^{-1}$) | $60.80 | 73 |
| BHT (50 mg $kg^{-1}$) + DHQ (10 mg $kg^{-1}$) | $10.70 | 86 |
| BHT (100 mg $kg^{-1}$) + DHQ (25 mg $kg^{-1}$) | $23.00 | 69 |

to be added in high concentration. This means that if these standards are being more costly than the alternative additives, it will strongly influence the final price of the biofuel.

Particularly, Roveda and coworkers as well as Comin and coworkers (Roveda et al. 2016; Comin et al. 2017) were responsible for showing that the addition of compounds containing the anthraquinone group offers greater stability to biodiesel when added to synthetic antioxidants, such as TBHQ, PG, and BHT. The authors have also shown that the use of ternary and binary combinations containing target alternative additives provides a highly satisfactory cost/benefit ratio (Table 3.1), since there is a reduction of up to approximately 88% in the cost of additives needed to provide higher stability and ensure that the biofuel meets quality specifications.

Roveda et al. (2016), for example, have shown that the mixtures containing PG (100 mg $kg^{-1}$) and the claimed alternative additive DHQ (50 mg $kg^{-1}$), when added to the soybean biodiesel, are responsible for an induction period (IP) of 13.7 h. Meanwhile, the same biodiesel containing only PG at a concentration of 500 mg $kg^{-1}$ showed an IP of 10 h. It is important to note that this combination not only allowed a significant reduction of 70% of the additives inserted into biodiesel, but also a reduction of 73% in the cost of additives. Likewise, Comin et al. (2017) showed that their proposed mixtures containing TBHQ, ALZ, and citric acid (CA) offered a reduction of cost (almost four times) with high benefit, assuming an average cost of about $43.60 USD per ton (combination), against $187.50 USD for additivation only with TBHQ. The mixture of additives containing 100 mg $kg^{-1}$ of TBHQ, 50 mg $kg^{-1}$ of ALZ, and 10 mg $kg^{-1}$of CA provided an IP of 14.6 h, while the mixture containing only TBHQ at 500 mg $kg^{-1}$ had the IP of 6.6 h. However, they extended their studies to show that when transition metals were added to the biodiesel, the combinations were potentially more suitable in maintaining the oxidative stability than the antioxidant TBHQ added individually at 500 mg $kg^{-1}$.

Thus, compounds containing the anthraquinone group are potentially suitable for biodiesel stabilization, especially in lowering the cost of additivation, since they can be used to combine with primary antioxidants at very low concentration. Even so, the mixtures are still effective to offer some remarkable improvement in biodiesel stability and all the combinations that have this type of compounds have been effective enough to ensure that the biofuel meets the quality specifications.

## References

Abbaszaadeh A, Ghobadian B, Omidkhah MR, Najafi G (2012) Current biodiesel production technologies: a comparative review. Energy Convers Manag 63:138–148

Adegoke GO, Vijay Kumar M, Gopala Krishna AG et al (1998) Antioxidants and lipid oxidation in foods: a critical appraisal. J Food Sci Technol 35:283–298

Agarwal AK, Khurana D (2013) Long-term storage oxidation stability of Karanja biodiesel with the use of antioxidants. Fuel Process Technol 106:447–452

Banga S, Varshney PK (2010) Effect of impurities on performance of biodiesel: a review

Barclay LRC, Vinqvist MR (2003) Phenols as Antioxidants. In: The chemistry of phenols, John Wiley & Sons, Ltd, pp 839–908

Canakci M, Sanli H (2008) Biodiesel production from various feedstocks and their effects on the fuel properties. J Ind Microbiol Biotechnol 35:431–441

Chahine MH, Macneill RF (1974) Effect of stabilization of crude whale oil with tertiary butylhydroquinone and other antioxidants upon keeping quality of resultant deodorized oil. A feasibility study. J Am Oil Chem Soc 51:37–41

Choe E, Min DB (2009) Mechanisms of antioxidants in the oxidation of foods. Compr Rev Food Sci Food Saf 8:345–358

Comin M, de Souza ACD, Roveda AC et al (2017) Alternatives binary and ternary blends and its effects on stability of soybean biodiesel contaminated with metals. Fuel 191:275–282. https://doi.org/10.1016/j.fuel.2016.11.080

Cosgrove JP, Church DF, Pryor WA (1987) The kinetics of the autoxidation of polyunsaturated fatty acids. Lipids 22:299–304

da Silva WLG, Salomão AA, Vila MMDC, Tubino M (2017) Influence of water and ultraviolet irradiation on the induction period of the oxidation of biodiesel. J Braz Chem Soc 28:676–680

Damasceno SS, Santos NA, Santos IMG et al (2013) Caffeic and ferulic acids: an investigation of the effect of antioxidants on the stability of soybean biodiesel during storage. Fuel 107: 641–646. https://doi.org/10.1016/j.fuel.2012.11.045

De Sousa LS, De Moura CVR, De Oliveira JE, De Moura EM (2014) Use of natural antioxidants in soybean biodiesel. Fuel 134:420–428. https://doi.org/10.1016/j.fuel.2014.06.007

de Souza ACD, Comin M, de Oliveira LH et al (2017) Application of solvent dye in the field of biodiesel preservation. Color Technol 133:165–169. https://doi.org/10.1111/cote.12263

Decker EA, Elias RJ, McClements DJ (2010) Oxidation in foods and beverages and antioxidant applications: management in different industry sectors, Elsevier

Deyab MA (2016) Corrosion inhibition of aluminum in biodiesel by ethanol extracts of Rosemary leaves. J Taiwan Inst Chem Eng 58:536–541

Diana da Silva Araújo F, Araújo IC, Costa ICG, et al (2014) Study of degumming process and evaluation of oxidative stability of methyl and ethyl biodiesel of Jatropha curcas L. oil from three different Brazilian states. Renew Energy. https://doi.org/10.1016/j.renene.2014.06.001

Dwivedi G, Sharma MP (2015) Effect of metal on stability and cold flow property of pongamia biodiesel. Mater Today Proc 2:1421–1426

Embuscado ME (2015) Herbs and spices as antioxidants for food preservation. Handb Antioxid Food Preserv 251–283

Fattah IMR, Masjuki HH, Kalam MA et al (2014) Effect of antioxidants on oxidation stability of biodiesel derived from vegetable and animal based feedstocks. Renew Sustain Energy Rev 30:356–370

Fiege H, Voges H, Hamamoto T et al (2000) Phenol derivatives

Grajzer M, Prescha A, Korzonek K et al (2015) Characteristics of rose hip (Rosa canina L.) cold-pressed oil and its oxidative stability studied by the differential scanning calorimetry method. Food Chem 188:459–466

Haas MJ, McAloon AJ, Yee WC, Foglia TA (2006) A process model to estimate biodiesel production costs. Bioresour Technol 97:671–678. https://doi.org/10.1016/j.biortech.2005.03.039

Håkansson B, Jägerstad M (1990) The effect of thermal inactivation of lipoxygenase on the stability of vitamin E in wheat. J Cereal Sci 12:177–185. https://doi.org/10.1016/S0733-5210(09)80099-4

Hraš AR, Hadolin M, Knez Ž, Bauman D (2000) Comparison of antioxidative and synergistic effects of rosemary extract with α-tocopherol, ascorbyl palmitate and citric acid in sunflower oil. Food Chem 71:229–233

Jain S, Sharma MP (2012) Application of thermogravimetric analysis for thermal stability of Jatropha curcas biodiesel. Fuel 93:252–257

Jakeria MR, Fazal MA, Haseeb A (2014) Influence of different factors on the stability of biodiesel: a review. Renew Sustain Energy Rev 30:154–163

Kivevele T, Huan Z (2015) Influence of metal contaminants and antioxidant additives on storage stability of biodiesel produced from non-edible oils of Eastern Africa origin (Croton megalocarpus and Moringa oleifera oils). Fuel 158:530–537

Knothe G (2005) Dependence of biodiesel fuel properties on the structure of fatty acid alkyl esters. Fuel Process Technol 86:1059–1070

Knothe G (2007) Some aspects of biodiesel oxidative stability. Fuel Process Technol 88:669–677

Knothe G, Razon LF (2017) Biodiesel fuels. Prog Energy Combust Sci 58:36–59. https://doi.org/10.1016/j.pecs.2016.08.001

Kreivaitis R, Gumbyte M, Kazancev K et al (2013) A comparison of pure and natural antioxidant modified rapeseed oil storage properties. Ind Crops Prod 43:511–516. https://doi.org/10.1016/j.indcrop.2012.07.071

Lamba BY, Joshi G, Tiwari AK et al (2013) Effect of antioxidants on physico-chemical properties of EURO-III HSD (high speed diesel) and Jatropha biodiesel blends. Energy 60:222–229

Lapuerta M, Rodríguez-Fernández J, Ramos Á, Álvarez B (2012) Effect of the test temperature and anti-oxidant addition on the oxidation stability of commercial biodiesel fuels. Fuel 93:391–396

Lima RP, da Luz PTS, Braga M et al (2017) Murumuru (Astrocaryum murumuru Mart.) butter and oils of buriti (Mauritia flexuosa Mart.) and pracaxi (Pentaclethra macroloba (Willd.) Kuntze) can be used for biodiesel production: Physico-chemical properties and thermal and kinetic studies. Ind Crops Prod 97:536–544

Liu G (2015) Development of low-temperature properties on biodiesel fuel: a review. Int J Energy Res 39:1295–1310

Mahmudul HM, Hagos FY, Mamat R et al (2017) Production, characterization and performance of biodiesel as an alternative fuel in diesel engines–A review. Renew Sustain Energy Rev 72:497–509

Maia ECR, Borsato D, Moreira I et al (2011) Study of the biodiesel B100 oxidative stability in mixture with antioxidants. Fuel Process Technol 92:1750–1755

Marinova E, Toneva A, Yanishlieva N (2008) Synergistic antioxidant effect of α-tocopherol and myricetin on the autoxidation of triacylglycerols of sunflower oil. Food Chem 106:628–633

Medeiros ML, Cordeiro AMMT, Queiroz N et al (2014) Efficient antioxidant formulations for use in biodiesel. Energy Fuels 28:1074–1080

Min DB, Smouse TH (1985) Flavor chemistry of fats and oils. The American oil chemists society

Mittelbach M, Schober S (2003) The influence of antioxidants on the oxidation stability of biodiesel. J Am Oil Chem Soc 80:817–823

Moser BR (2008) Efficacy of myricetin as an antioxidant in methyl esters of soybean oil. Eur J Lipid Sci Technol 110:1167–1174. https://doi.org/10.1002/ejlt.200800145

Moser BR (2012) Efficacy of gossypol as an antioxidant additive in biodiesel. Renew Energy 40:65–70. https://doi.org/10.1016/j.renene.2011.09.022

Niki E, Saito T, Kawakami A, Kamiya Y (1984) Inhibition of oxidation of methyl linoleate in solution by vitamin E and vitamin C. J Biol Chem 259:4177–4182

Nivetha S, Roy DV (2013) Effect of natural and synthetic antioxidants on oxidative stability of FAMEs obtained from hevea brasiliensis. J Energy Chem 22:935–941. https://doi.org/10.1016/S2095-4956(14)60275-8

Omura K (1995) Antioxidant synergism between butylated hydroxyanisole and butylated hydroxytoluene. J Am Oil Chem Soc 72:1565–1570

Orives JR, Galvan D, Pereira JL et al (2014) Experimental design applied for cost and efficiency of antioxidants in biodiesel. J Am Oil Chem Soc 91:1805–1811

Othman MF, Adam A, Najafi G, Mamat R (2017) Green fuel as alternative fuel for diesel engine: a review. Renew Sustain Energy Rev 80:694–709

Pantoja SS, da Conceição LRV, da Costa CEF et al (2013) Oxidative stability of biodiesels produced from vegetable oils having different degrees of unsaturation. Energy Convers Manag 74:293–298

Pisoschi AM, Pop A (2015) The role of antioxidants in the chemistry of oxidative stress: a review. Eur J Med Chem 97:55–74

Pokorny J, Yanishlieva N, Gordon MH (2001) Antioxidants in food: practical applications. CRC press

Pullen J, Saeed K (2012) An overview of biodiesel oxidation stability. Renew Sustain Energy Rev 16:5924–5950

Pullen J, Saeed K (2014) Experimental study of the factors affecting the oxidation stability of biodiesel FAME fuels. Fuel Process Technol 125:223–235

Rasimoglu N, Temur H (2014) Cold flow properties of biodiesel obtained from corn oil. Energy 68:57–60

Rawat DS, Joshi G, Lamba BY et al (2015) The effect of binary antioxidant proportions on antioxidant synergy and oxidation stability of Jatropha and Karanja biodiesels. Energy 84: 643–655

Richard-Forget F, Gauillard F, Hugues M et al (1995) Inhibition of horse bean and germinated barley lipoxygenases by some phenolic compounds. J Food Sci 60:1325–1329

Romanini DC, Trindade MAG, Zanoni MVB (2009) A simple electroanalytical method for the analysis of the dye solvent orange 7 in fuel ethanol. Fuel. https://doi.org/10.1016/j.fuel.2008.07.022

Roveda AC, Comin M, Caires ARL et al (2016) Thermal stability enhancement of biodiesel induced by a synergistic effect between conventional antioxidants and an alternative additive. Energy 109:260–265. https://doi.org/10.1016/j.energy.2016.04.111

Roveda AC, Tenório KV, Caires ARL et al (2017) Alternative binary blends and their synergistic effect on stability of soybean biodiesel. Clean Technol Environ Policy. https://doi.org/10.1007/s10098-017-1445-7

Rudnick LR (2013) Synthetics, mineral oils, and bio-based lubricants: chemistry and technology. CRC press

Sarin A, Arora R, Singh NP et al (2009) Influence of metal contaminants on oxidation stability of Jatropha biodiesel. Energy 34:1271–1275

Sarin A, Arora R, Singh NP et al (2010a) Effect of metal contaminants and antioxidants on the oxidation stability of the methyl ester of Pongamia. J Am Oil Chem Soc 87:567–572

Sarin A, Arora R, Singh NP et al (2010b) Synergistic effect of metal deactivator and antioxidant on oxidation stability of metal contaminated Jatropha biodiesel. Energy 35:2333–2337

Serqueira DS, Dornellas RM, Silva LG et al (2015) Tetrahydrocurcuminoids as potential antioxidants for biodiesels. Fuel 160:490–494. https://doi.org/10.1016/j.fuel.2015.07.104

Serrano M, Bouaid A, Martínez M, Aracil J (2013) Oxidation stability of biodiesel from different feedstocks: influence of commercial additives and purification step. Fuel 113:50–58. https://doi.org/10.1016/j.fuel.2013.05.078

Smyk B (2015) Singlet oxygen autoxidation of vegetable oils: Evidences for lack of synergy between β-carotene and tocopherols. Food Chem 182:209–216

Spacino KR, da Silva ET, Angilelli KG et al (2016) Relative protection factor optimisation of natural antioxidants in biodiesel B100. Ind Crops Prod 80:109–114. https://doi.org/10.1016/j.indcrop.2015.11.034

Sulistyo H, Almeida MF, Dias JM (2015) Influence of synthetic antioxidants on the oxidation stability of biodiesel produced from acid raw Jatropha curcas oil. Fuel Process Technol 132:133–138

Trindade MAG, Zanoni MVB (2009) Voltammetric sensing of the fuel dye marker Solvent Blue 14 by screen-printed electrodes. Sensors Actuators B Chem 138:257–263. https://doi.org/10.1016/j.snb.2009.01.043

Trindade MAG, Ferreira VS, Zanoni MVB (2007) A square-wave voltammetric method for analysing the colour marker quinizarine in petrol and diesel fuels. Dye Pigment. https://doi.org/10.1016/j.dyepig.2006.03.020

Trindade MAG, Zanoni MVB, Matysik F-M (2010) Sensitive determination of water insoluble dyes used as marking of commercial petroleum products using high-performance liquid chromatography with electrochemical detection. Electroanalysis. https://doi.org/10.1002/elan.200900500

Trindade MAG, Stradiotto NR, Zanoni MVB (2011) Corantes marcadores de combustíveis: legislação e métodos analíticos para detecção. Química Nov 34:1683–1691

Trindade MAG, Romanini DC, Zanoni MVB (2012) Determination of Sudan II dye in ethanol fuel by chromatographic and electroanalytical methods

Trindade MAG, Bilibio U, Zanoni MVB (2014) Enhancement of voltammetric determination of quinizarine based on the adsorption at surfactant-adsorbed-layer in disposable electrodes. Fuel 136:201–207. https://doi.org/10.1016/j.fuel.2014.07.044

Varatharajan K, Pushparani DS (2017) Screening of antioxidant additives for biodiesel fuels. Renew Sustain Energy Rev. https://doi.org/10.1016/j.rser.2017.07.020

Yehye WA, Rahman NA, Ariffin A et al (2015) Understanding the chemistry behind the antioxidant activities of butylated hydroxytoluene (BHT): a review. Eur J Med Chem 101:295–312

# Chapter 4
# Hydroxychalcones: Synthetic Alternatives to Enhance Oxidative Stability of Biodiesel

**Caroline Carapina Da Silva, Bruna Silveira Pacheco, Samantha Coelho de Freitas, Lucas Moraes Berneira, Marco Aurélio Ziemann dos Santos, Lucas Pizzuti and Claudio Martin Pereira de Pereira**

**Abstract** In this chapter, the main aspects about biodiesel synthesis and established quality parameters are reviewed and discussed in the perspective of oxidative stability. First, the characteristics, which define biodiesel as a promising, renewable green fuel, are listed. On the other hand, the propensity of biodiesel to oxidation is explained as one of the limitations to its application. Along this line of thought, the majority of natural and synthetic antioxidants usually applied in the industry are discussed, in which it can be noted that phenolic compounds are very representative. Chalcones containing hydroxyl groups in its structure are phenolic compounds, and so, their origins, sources, and synthetic approaches to obtain these types of molecules are presented. Antioxidant effects of hydroxychalcones already reported in the literature are pointed, and then an experimental topic shows the influence of two synthetic hydroxychalcones, namely (*E*)-1-(2-hydroxyphenyl)-3-phenylprop-2-en-1-one and (*E*)-1-(2-hydroxyphenyl)-3-(3-hydroxyphenyl) prop-2-en-1-one, on the oxidative stability of biodiesel obtained from frying soybean oil. The effects assessed by calorimetric analysis revealed interesting effects on the enhancement of biodiesel oxidative stability.

**Keywords** Biodiesel · Oxidative stability · Antioxidant · Hydroxychalcone Synthetic chalcones

C. C. Da Silva · B. S. Pacheco · S. C. de Freitas · L. M. Berneira
M. A. Z. dos Santos · C. M. P. de Pereira (✉)
Laboratory of Lipidomics and Bioorganic Center of Chemical, Pharmaceutical and Food Sciences, Federal University of Pelotas, Pelotas, Rio Grande do Sul 96001-970, Brazil
e-mail: claudiochemistry@gmail.com; lahbbioufpel@gmail.com

L. Pizzuti
Faculty of Exact Sciences and Technology, Federal University of Grande Dourados, Highway Dourados-Itahum, km 12, PO Box 364, Dourados, MS 79804-970, Brazil

 
M. Trindade (ed.), *Increased Biodiesel Efficiency*, Green Energy and Technology,
https://doi.org/10.1007/978-3-319-73552-8_4

## 4.1 Biodiesel

Biodiesel is a combustible chemically constituted of a mixture of several alkyl esters of fatty acids. This product can be obtained from vegetables, animals, microorganisms, and waste materials sources (Rockembach et al. 2014; Vieira et al. 2016; Marcon et al. 2017). These biological matrices make biodiesel a green alternative to fossil fuels since it is less aggressive to the environment, contributing to the recycling of resources and generating less pollution (Sajjadi et al. 2016). Moreover, it is known that the reserves of fossil fuels will be depleted in the near future although the energy demand continues to increase. In this way, the development of new sources of power, such as biodiesel, is considerably important for the sustainability of future generations (De Oliveira et al. 2013; Knothe and Razon 2017).

Although biodiesel is considered a green fuel for having several advantages, such as being biodegradable, non-toxic, environmental-friendly, having low sulfur content and being renewable, it faces numerous limitations in order to be fully implemented (Mekhilef et al. 2011). Major limitations are the higher cost of production compared to conventional fuels and the possible deviation of food resources to produce energy (Singh et al. 2011). Besides, due to its chemical constitution, biodiesel is susceptible to oxidation and has some other properties drawbacks involving cold flow and viscosity (Hoekman et al. 2012).

Generally, biodiesel is synthesized throughout a transesterification reaction, which converts triglycerides from the fatty matrix to alkyl esters of fatty acids. This process is accomplished employing methanol or another short-chain alcohol as a reactant and alkali or acid as catalysts (Pereira et al. 2012). The final product of this reaction—biodiesel—is more suitable to be used as a combustible than the crude oil since its viscosity, density, and volatility are more adequate for combustion in engines. Moreover, prejudicial factors that could happen when a crude oil is combusted, such as gumming and formation of carbon deposits, are avoided in this conversion procedure (Singh and Singh 2010).

In order to be used, the characteristics of the biodiesel produced must agree with the minimum standards set by Europe (EN 14214) or by the USA (ASTM D6751) (Knothe 2006). Both these requirements serve as models worldwide and are related to several properties of the fuel, such as kinematic viscosity, iodine value, cetane number, acid value, mineral content, and oxidation stability. Ideally, a certain synthesized biodiesel should be in line with the values established for these parameters. In most of the cases, however, it fails in one or more of the parameters, mainly due to its fatty acid composition (Pinzi et al. 2009).

Nowadays, edible oils such as soybean, rapeseed, palm, and sunflower oil are the main sources for the production of biodiesel. However, alternative feedstock like non-edible and waste oils are gradually growing in the market, since 75% of biodiesel production costs are related to its source (Gui et al. 2008). These alternative oils also have other advantages compared to conventional feedstock, since they are not used in the food market and may occur abundantly from region to region (Sajjadi et al. 2016; Knothe and Razon 2017).

In many countries, the production of biodiesel has been promoted or facilitated through regulatory efforts, in which the top producer is the USA (4.7 billion liters). Overall, other countries with a considerable production of biodiesel include Brazil, Germany, Argentina, Indonesia, and France (Knothe and Razon 2017). In Brazil, the synthesis of this fuel was approximately 3.87 billion liters in 2016, with the majority of its feedstock being from soybean oil (68%). Figure 4.1 shows other sources that include animal tallow (14%) and waste cooking oil (2.5%).

As mentioned above, the use of waste oils is promising starting materials for the production of biodiesel. These types of oils can be reutilized, thus returning to the energetic chain and minimizing possible environmental impacts (Kulkarni and Dalai 2006). On the other hand, some barriers prevent the use of these oils, mainly due to lack of infrastructure and logistics for their efficient harvest. Moreover, most waste oils need to overcome a previous cleaning process in order to be transesterified, which could increase its production cost, making it difficult for an extensive implementation (Wang et al. 2007; Lapuerta et al. 2008).

Generally, the fatty acid profile of most vegetable oils is constituted of common biomolecules such as palmitic, stearic, oleic, linoleic, and linolenic acid with concentrations that vary from matrix to matrix (Atabani et al. 2013). These variations are linked directly to the properties of biodiesel including oxidative stability. A higher amount of unsaturated fatty acids leads to an oxidative instability in the fuel as the double bonds present in the fatty acids are reactive toward oxidation reactions (Knothe 2005).

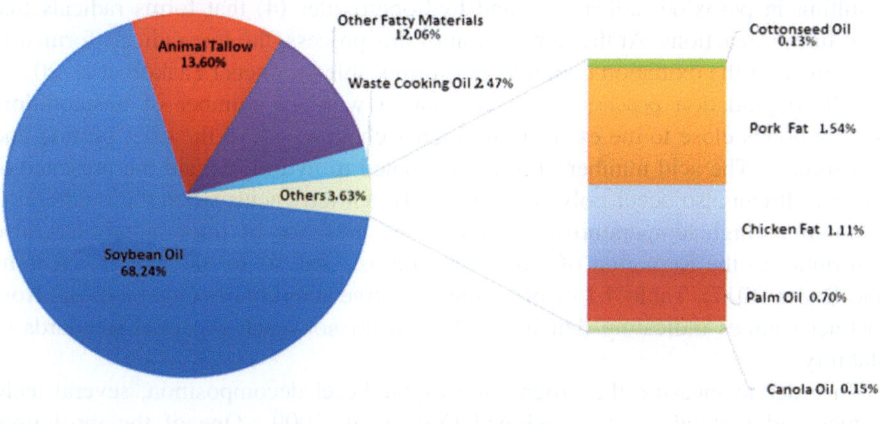


**Fig. 4.1** Brazilian feedstock matrix of biodiesel (Pinto 2017)

### 4.1.1 Oxidative Stability

Oxidative stability can be defined as the amount of time required to oxidize a certain portion of biodiesel. This subject is extensively researched since it affects the fuel in early stages of use, during transportation, and storage (Knothe 2007). The presence of unsaturation is related to autoxidation, which is influenced by the presence of air, heat, metals, peroxides, and light, among other factors. This important parameter is evaluated by ASTM D6751 and EN 14214 that indicate a minimum induction time of 3 and 8 h, respectively (Dantas et al. 2011).

The oxidation of biodiesel influences greatly in the quality of the fuel. Its degradation is linked to variations in others physicochemical properties such as acid number, kinematic viscosity, polymer content, iodine value, and peroxide value. In this context, the oxidized biodiesel can lead to several problems in the operational system of a vehicle, causing corrosion, obstruction, and formation of deposits in the interior of the engine (Kumar 2017).

The oxidation process in biodiesel that leads to its deterioration is very complex and involves the development of hydroperoxides. These reactive species initiate several secondary reactions, which result in the formation of carboxylic acids, ketones, aldehydes, polymers, and other undesired compounds. In general, the degradation is composed of a free radical process divided in stages, namely initiation, propagation, and termination, which are outlined in Fig. 4.2 (Fu et al. 2017).

As it can be seen in Fig. 4.2, the oxidation undergoes a primary reaction in which hydrogen is removed of a polyunsaturated fatty acid (1). This step is relatively slow and forms the first free radicals (2) that initiate the degradation of the biodiesel. The propagation stage is characterized by the reaction with oxygen, resulting in peroxide radicals (3) and hydroperoxides (4) that forms radicals that feed further reactions. At the termination of this process, the free radicals formed in the course of the oxidation combine and form stable products (Yaakob et al. 2014).

The degradation process has a correlation with the number of unsaturation, double bonds close to the ester group, higher chain length of the ester portion and *cis*-isomers. The acid number of a fuel is related to hydrolysis and the presence of oxygen. It can also occur polymerization between fatty acids when the biodiesel is exposed to high temperature conditions. The presence of trace metals can also contribute to the formation of radicals acting as catalysts to this process (Knothe and Razon 2017). Table 4.1 displays the oxidative stability of some biodiesel from distinct sources indicating that most of them do not reach accepted standards of stability.

In order to measure the progression of biodiesel decomposition, several techniques and methods were developed (Xin et al. 2009). One of the most used methods is the Rancimat, in which a stream of air is bubbled into the sample at elevated temperatures. Other methods can be cited, for example the iodine value, the active oxygen bomb test, and Schaal's oven test. It must be noted that thermal analysis is emerging in this field, mainly by differential scanning calorimetry (DSC) and differential thermal analysis (Herbinet et al. 2010; Saluja et al. 2016).

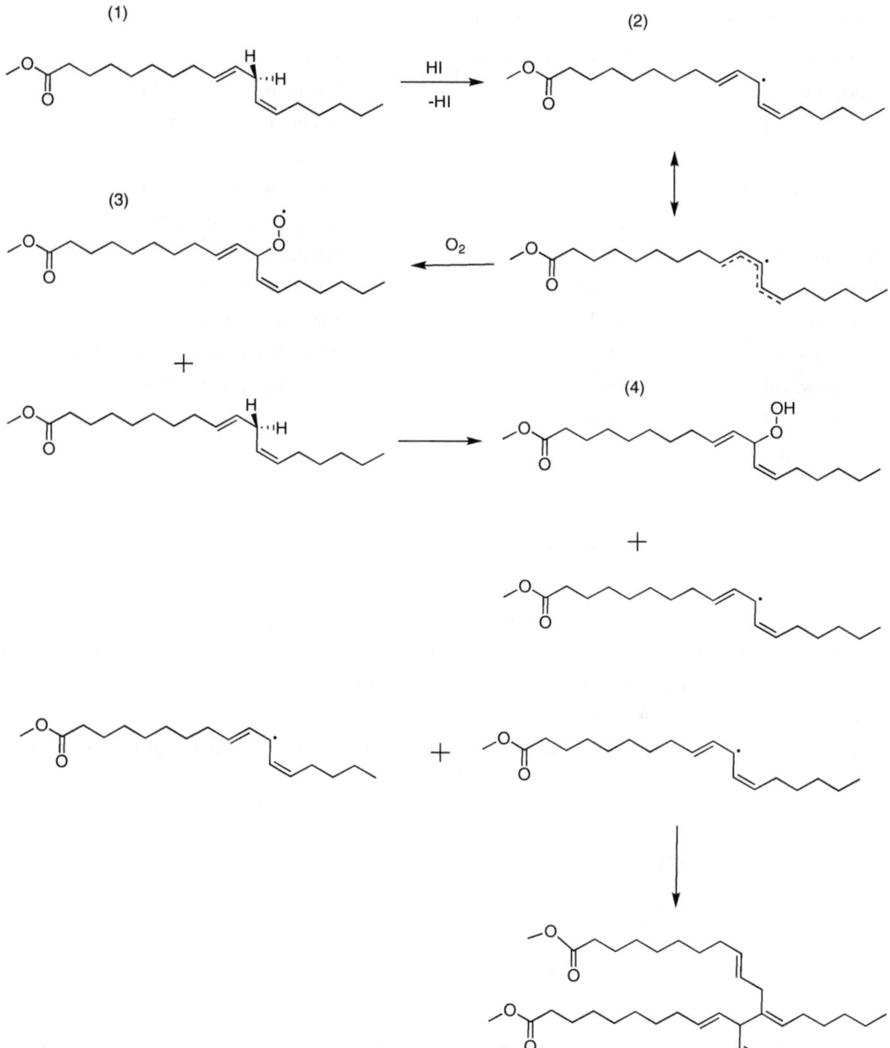


**Fig. 4.2** Mechanism of linoleic acid methyl ester oxidation

DSC is a thermal technique that measures energetic differences between the sample and a reference material as a function of temperature. It has been extensively used to characterize several physical and chemical properties of combustibles, lubricant, oils, polymers, and other materials (Dunn 2006). In this technique, a small quantity of the sample is required and it is contained in a crucible. Then, the sample alongside a reference material is placed in a furnace in which the temperature and atmosphere can be controlled and measured, resulting in endothermic or exothermic events (Stuart 2012).

**Table 4.1** Biodiesel and its respective oxidative stability

| Biodiesel source | Oxidative stability (h) | References |
|---|---|---|
| Corn oil | 6.5 | Serqueira et al. (2014) |
| Sunflower oil | 1.73 | Sarin et al. (2007) |
| Soybean oil | 3.80 | Sarin et al. (2007) |
| Waste cooking oil | 2.6 | Serqueira et al. (2014) |
| Colza oil | 11.2 | Serqueira et al. (2014) |
| Karanjia oil | 2.35 | Sarin et al. (2007) |
| Jatropha oil | 3.23 | Sarin et al. (2007) |
| Palm oil | 13.37 | Sarin et al. (2007) |
| Beef tallow | 1.2 | Mittelbach and Schober (2003) |
| Rapeseed oil | 9.1 | Mittelbach and Schober (2003) |
| Safflower oil | 0.86 | Xin et al. (2009) |
| Cottonseed oil | 3.6 | Serqueira et al. (2014) |

### *4.1.2 Antioxidants in Biodiesel*

It has been established that susceptibility to oxidation is one of the major drawbacks of biodiesel usage. In this way, it is very important to combine biodiesel with antioxidants that are effective for the minimization of oxidation stability issues (Yaakob et al. 2014). Antioxidants may be isolated from natural sources, serving as examples α-tocopherol (vitamin E), citric acid, and caffeic acid (Fig. 4.3).

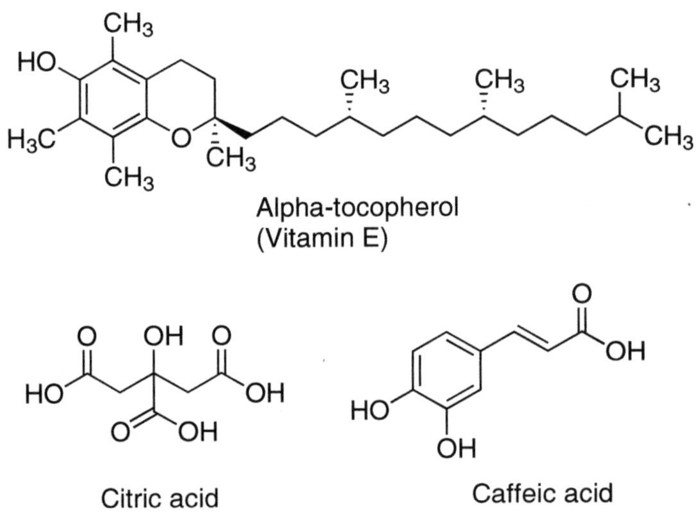


**Fig. 4.3** Structure of some naturally derived biodiesel antioxidants: vitamin E, citric acid, and caffeic acid (Yaakob et al. 2014)

Vegetable oils may have considerate quantities of natural antioxidants, but they are reduced during oil refining processes.

Antioxidants derived from synthetic sources are generally preferable and more effective, in which the major examples include, but are not limited to: butylated hydroxytoluene (BHT), butylated hydroxyanisole (BHA), *tert*-butylated hydroxyquinone (TBHQ), propyl 3,4,5-trihydroxybenzoate (PG), 1,2,3-trihydroxybenzene (pyrogallol—PY), and gallic acid (GA) (Fig. 4.4). Depending on their mode of action, they can be classified as primary antioxidants that are free radical terminators, metal ion chelators, or oxygen scavengers. Secondary antioxidants are preventive and can act as chain breakers (peroxide radical quenchers), or reducing agents (hydroperoxide decomposers), retarding the chain initiation reaction (Shahidi and Wanasundara 1992; Yaakob et al. 2014).

These antioxidants or potentially new ones are blended with biodiesel in relatively low concentrations ranging from 200 to 2000 ppm (1 ppm: parts per million = 1 μg/mL = 1 mg/L) to minimize added costs (Dunn 2008; Pullen and Saeed 2012). Effectiveness of antioxidants will vary depending on the composition of fatty acid methyl esters, feedstock, and oxidative stability of the biodiesel (Rizwanul Fattah et al. 2014b; Yaakob et al. 2014).

It can be seen that antioxidant agents are diverse in molecular structure, but most of them are phenolic substances. Phenolic compounds tend to have a proximate relationship with antioxidant activity, because the hydrogen from hydroxyl groups can be easily donated, forming relatively stable radical intermediates (Cao et al. 1997; Heim et al. 2002; Senevirathne et al. 2006; Detsi et al. 2009). The –OH groups scavenge reactive radicals, such as peroxyl radicals, and can be stabilized by

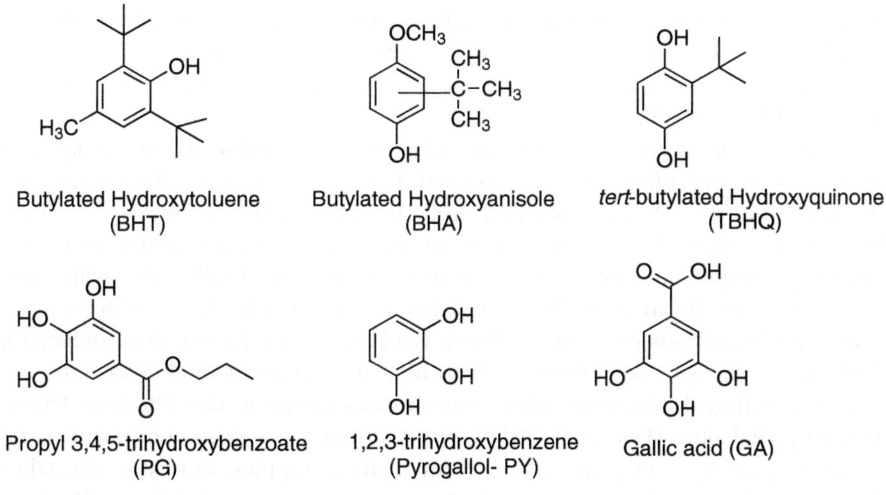


**Fig. 4.4** Structure of some synthetic antioxidants used for biodiesel storage (Rizwanul Fattah et al. 2014b; Yaakob et al. 2014)

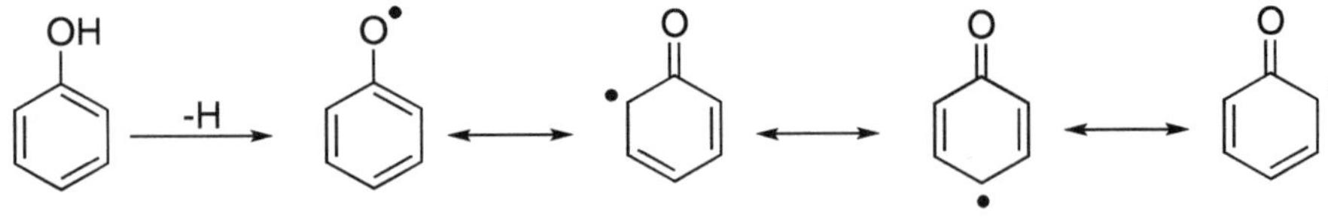


**Scheme 4.1** Stabilization of phenoxy radical by delocalization of unpaired electrons around the aromatic ring (Rizwanul Fattah et al. 2014b)

resonance delocalization and lack of apt sites to be attacked by oxygen (Scheme 4.1) (Rizwanul Fattah et al. 2014a, b).

Phenolic groups are frequently present in chalcones, which are molecules naturally abundant in edible plants or obtained synthetically. Chalcones are very versatile molecules and have been associated with many biological activities, in which antioxidant activity can be highlighted (Detsi et al. 2009; Righi et al. 2011; Serifi et al. 2013; Días-Tielas et al. 2016).

## 4.2 Chalcones and Antioxidant Activity

Chalcones (1, 3-diaryl-2-propen-1-ones) are a class of molecules within the family of flavonoids, which present two aromatic rings connected by a three-carbon linker of an $\alpha$-$\beta$-unsaturated system. As it can be seen from Fig. 4.5, one of the aromatic rings (Ring "A") is connected to a carbonyl group, while the other aromatic ring (Ring "B") is connected to the $\alpha$-$\beta$-unsaturation (Dawane et al. 2009).

Chalcones are a category of open-chain flavonoids, and in fact, some of them are immediate precursors of other classes of flavonoids, such as aurones. It can be seen from Fig. 4.6 that chalcones have a close structural and biogenetic relationship with flavonoids, including aurones, and so they can co-occur in many natural products (Veith and Grayer 2006).

Chalcones and aurones are major contributors to the yellow to orange pigments seen in flowering plants, mainly isolated from Leguminosae, Asteraceae, and Moraceae families (Rozmer and Perjési 2016). These phytochemicals are widely distributed in seeds, leaves, and flowers of plants and are associated to protection against photosynthetic stress, reactive oxygen species (ROS), and pathogenic microorganisms (Serifi et al. 2013). The key enzyme to chalcone synthesis in the flavonoid biosynthesis pathway, chalcone synthase, is also known to be induced in plants under stressful conditions, such as high irradiation of ultraviolet light, and bacterial, or fungal infections, which corroborates the protective effective demonstrated by chalcones (Dao et al. 2011). The distribution of these compounds is not limited to plants, as they are also present in fruits (apples, citruses), vegetables (tomatoes, potatoes), beer, wine, and other dietary components (Tomás-Barberán and Clifford 2000; Orlikova et al. 2011).

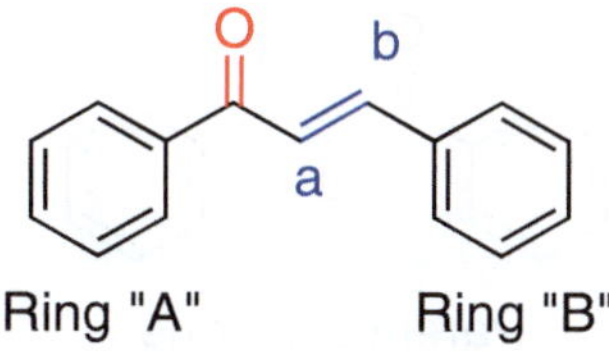


**Fig. 4.5** General structure of a chalcone (Detsi et al. 2009)

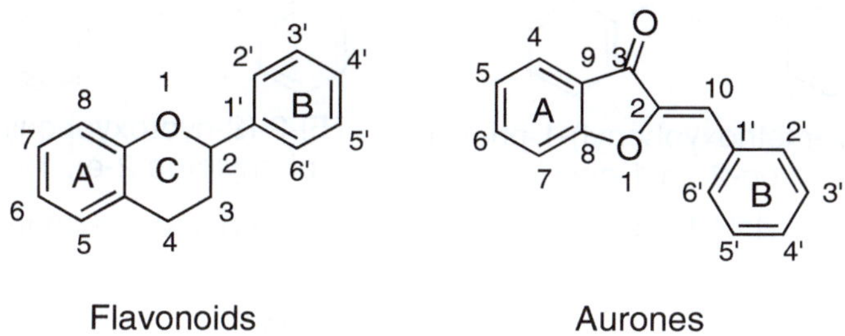


**Fig. 4.6** General structure of flavonoids (Heim et al. 2002) and aurones (Detsi et al. 2009)

Apart from the $\alpha$-$\beta$-unsaturation, the general structure of chalcones is highly versatile because the aromatic rings can present a wide range of substituents in any position. Thus, many molecules of natural or synthetic chalcones can be isolated or obtained, which will differ structurally depending on the substitution pattern of the parent chalcone (Orlikova et al. 2011; Ritter et al. 2015). Possibilities include, but are not limited to, methoxylated, hydroxylated, halogenated, and *N*-containing chalcones (Padhye et al. 2010). Simple examples of this classification can be seen in Fig. 4.7.

Chalcones can be multifunctional molecules since one particular compound may exhibit more than one biological activity, such as the antioxidant, anti-inflammatory, and chemopreventive activities associated with licochalcone (Fig. 4.8), which is present in licorice (Lee et al. 2008; Franceschelli et al. 2011).

Many studies have reported an antioxidant activity exhibited by chalcones and chalcone derivatives, especially chalcones containing hydroxyl groups (hydroxychalcones) (Mathiesena et al. 1996; Heim et al. 2002; Rezk et al. 2002; Sikander et al. 2011). The radical-scavenging activity of chalcones is influenced by the aryl groups and their substituents (rings “A” and “B” in Fig. 4.5). It is believed that the presence of one or more hydroxyl groups can be key contributors to the antioxidant effect associated with chalcones (Mathew et al. 2017).

Sikander et al. (2011) showed that trans-chalcone (labeled as compound 1 in Fig. 4.9) inhibited reactive oxygen-triggered lesions induced by hydrogen peroxide ($H_2O_2$), resulting in the protection of hepatocellular carcinoma cells.

Rezk et al. (2002) showed the potent antioxidant activity of a dihydrochalcone, phloretin (2) and of its respective glucoside, phloridzin (phloretin 2-b-D-glucose)

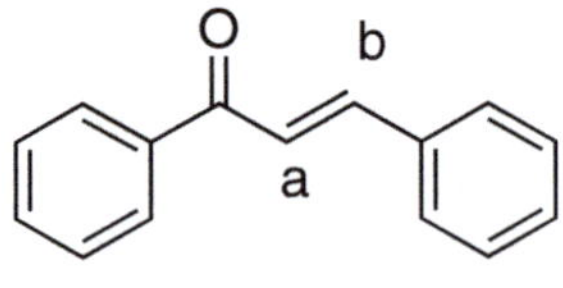


"Parent chalcone"

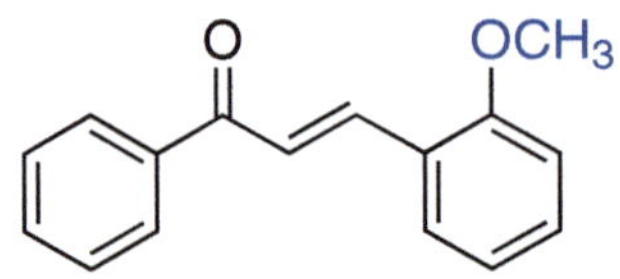


(*E*)-3-(2-methoxyphenyl)-1-phenyl prop-2-en-1-one

methoxylated chalcone

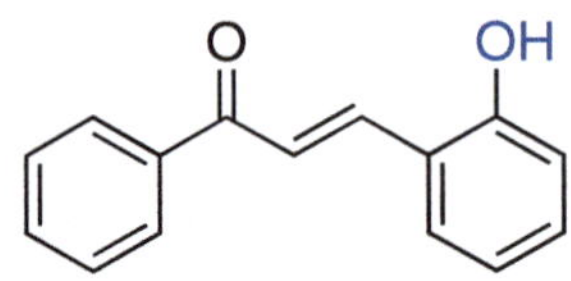


(*E*)-3-(2-hydroxyphenyl)-1-phenylprop-2-en-1-one

hydroxylated chalcone

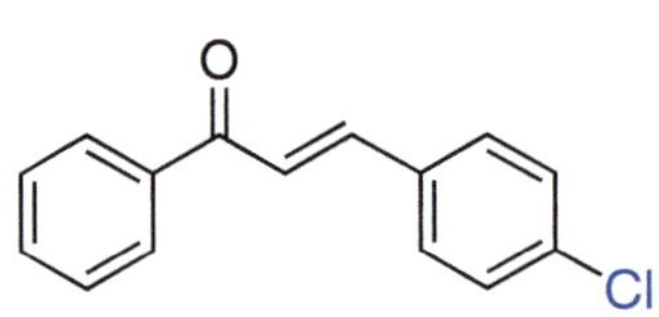


(*E*)-3-(4-chlorophenyl)-1-phenyl prop-2-en-1-one

halogenated chalcone

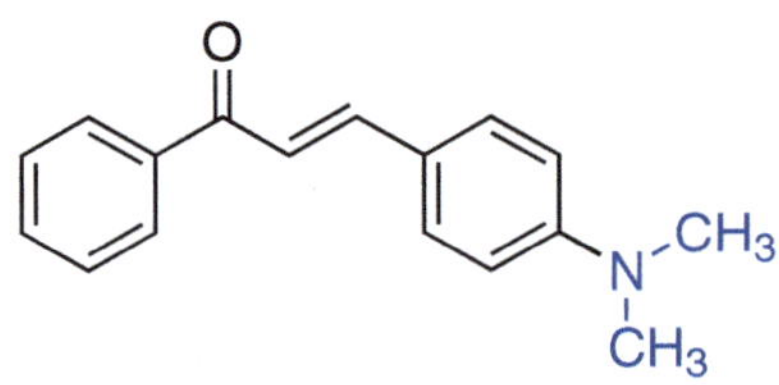


(*E*)-3-(4-(dimethylamino)phenyl)-1-phenylprop-2-en-1-one

N-containing chalcone

**Fig. 4.7** Examples of different chalcones based on their substitution pattern

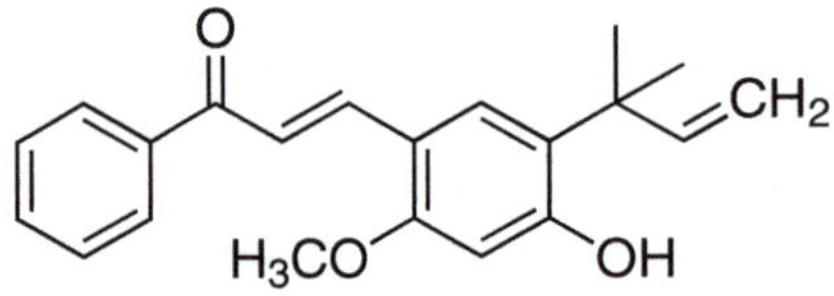


Licochalcone-A

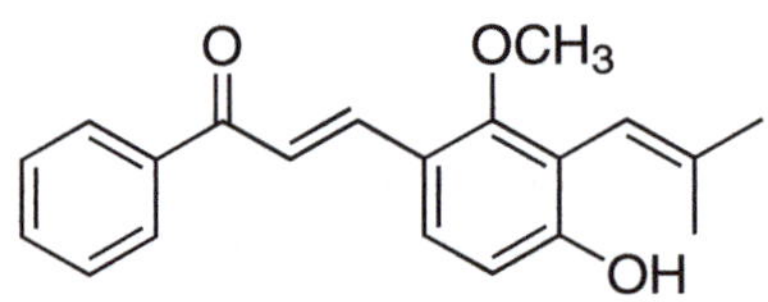


Licochalcone-C

**Fig. 4.8** Molecular structures of licochalcone-A and licochalcone-C associated with antioxidant, anti-inflammatory, and chemopreventive activities

(3), which are chalcones without $\alpha$-$\beta$ double bond containing four and three hydroxyl groups, respectively (Fig. 4.10). The concentration needed to scavenge 50% of peroxynitrite ($IC_{50}$ PON) and the concentration needed to inhibit 50% of lipid peroxidation ($IC_{50}$ LOP) were determined. For phloretin, $IC_{50}$ PON was

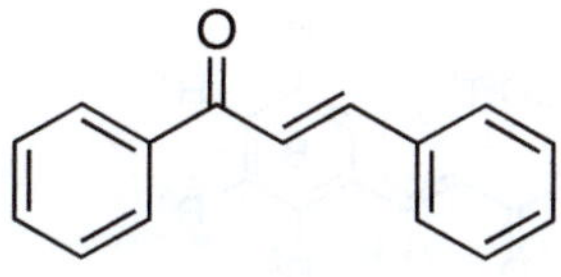


Compound **1**
Trans chalcone

**Fig. 4.9** Structure of trans-chalcone (compound 1) (Sikander et al. 2011)

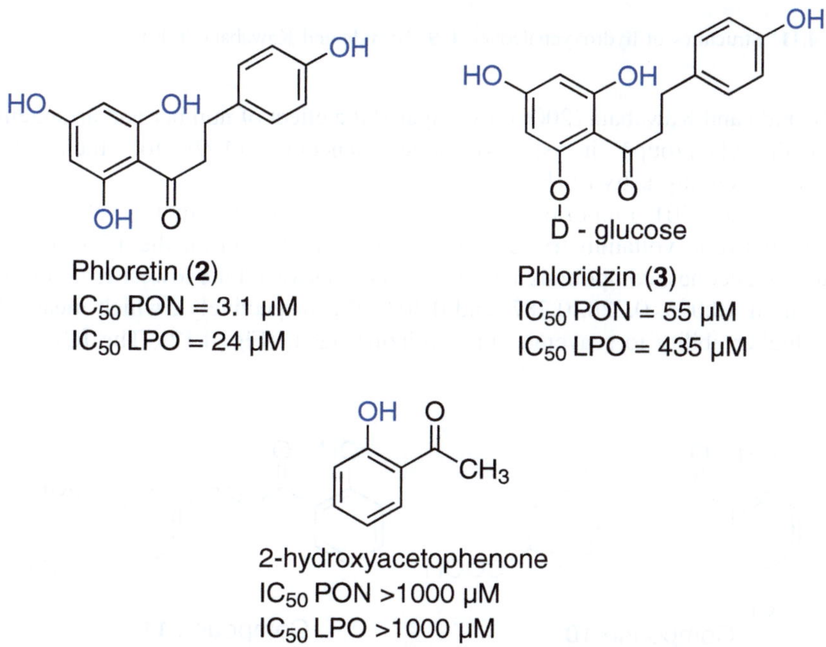


**Fig. 4.10** Structures of phloretin, phloridzin, 2-hydroxyacetophenone, and concentrations needed to scavenge 50% of peroxynitrite ($IC_{50}$ PON) and inhibit lipid peroxidation ($IC_{50}$ LOP) (Rezk et al. 2002)

3.1 μM and $IC_{50}$ LOP was 24 μM, while phloridzin's $IC_{50}$ PON was 55 μM and $IC_{50}$ LOP was 435 μM. The significance of these values can be understood when the authors also determined these values for a synthetic precursor of chalcones, 2-hydroxyacetophenone, whose $IC_{50}$ values were higher than 1000 μM. The authors proposed that the antioxidant pharmacophore in phloretin is the 2,6-dihydroxyacetophenone group due to its stabilization of radicals via tautomerization.

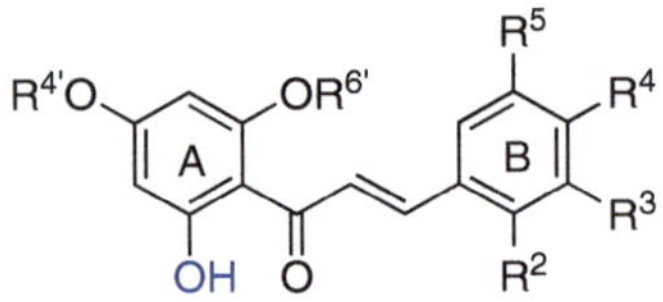

**4** $R^2 = R^5 = H$, $R^3 = R^4 = OH$, $R^{4'} = R^{6'} = CH_3$ (2',3,4-trihydroxy-4',6'-dimethoxychalcone)
**5** $R^2 = R^5 = R^{4'} = R^{6'} = H$, $R^3 = R^4 = OH$ (2',3,4,4',6'-pentahydroxychalcone)
**6** $R^4 = R^5 = H$, $R^2 = R^3 = OH$, $R^{4'} = R^{6'} = CH_3$ (2,2',3-trihydroxy-4',6'-dimethoxychalcone)
**7** $R^4 = R^5 = R^{4'} = R^{6'} = H$, $R^2 = R^3 = OH$ (2,2',3,4',6'-pentahydroxychalcone)
**8** $R^1 = H$, $R^3 = R^4 = R^5 = OH$, $R^{4'} = R^{6'} = CH_3$ (2',3,4,5-tetrahydroxy-4',6'-dimethoxychalcone)
**9** $R^1 = R^{4'} = R^{6'} = H$, $R^3 = R^4 = R^5 = OH$ (2',3,4,4',5,6'-hexahydroxychalcone)

**Fig. 4.11** Structures of hydroxychalcones 4–9 (Nishida and Kawabata 2006)

Nishida and Kawabata (2006) investigated the effect of number and arrangement of hydroxyl groups in six synthetic chalcones (4–9) for their DPPH radical-scavenging activity (Fig. 4.11).

Serifi et al. (2013) reported an antioxidant behavior of 2′-hydroxychalcones by means of cyclic voltammetry and chronoamperometry. From the 13 synthesized hydroxychalcones, compounds 10, 11, 12, and 13 showed the lowest $IC_{50}$ for $H_2O_2$ scavenging (0.095, 0.228, 0.287, and 0.303 mM, respectively), which means they have higher ability to scavenge this oxidizing agent (Fig. 4.12). The $IC_{50}$ in this

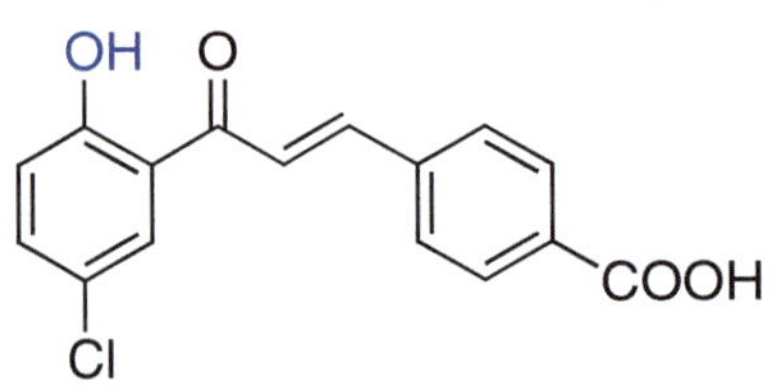


Compound **10**
$IC_{50}$ =0.095

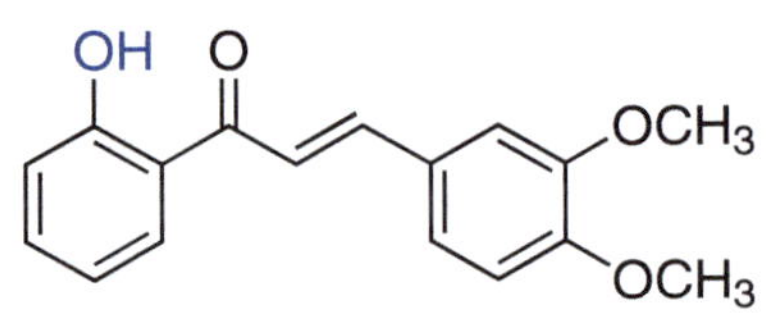


Compound **11**
$IC_{50}$ = 0.228

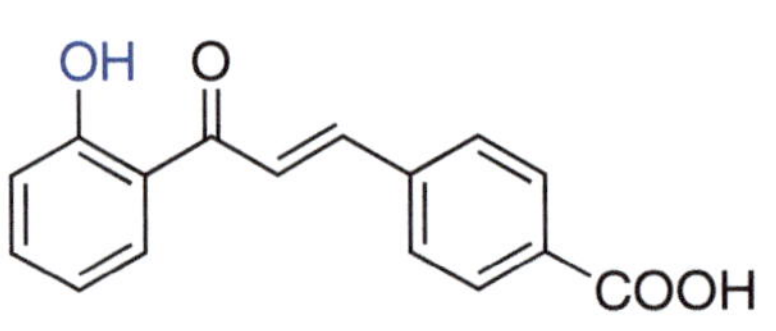


Compound **12**
$IC_{50}$ = 0.287

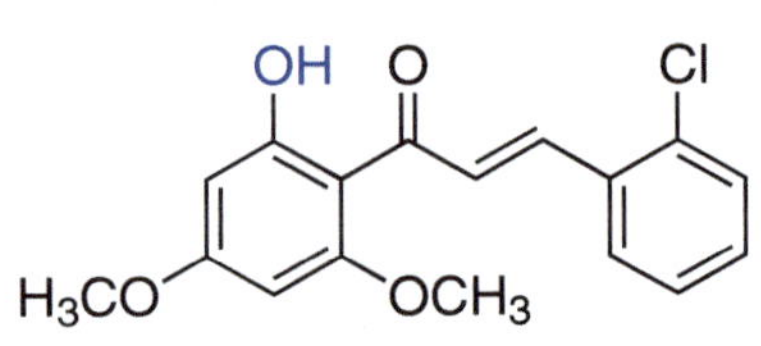


Compound **13**
$IC_{50}$ = 0.303

**Fig. 4.12** Structures of 2′-hydroxychalcones (10–13) with reported antioxidant behavior; 2′ hydroxyl groups are highlighted in blue; $IC_{50}$ is expressed in mM as the concentration of the compound required to scavenge 50% of $H_2O_2$ radicals (Serifi et al. 2013)

case is the concentration of the chalcone required to scavenge 50% of radicals (Senevirathne et al. 2006).

Besides experimental studies, some theoretical studies have been made in order to better understand the structure–activity relationship (SAR) on the antioxidant activity of hydroxychalcones (Xue et al. 2012; Mathew et al. 2017; Niu et al. 2017). Two main mechanisms by which hydroxychalcones can act as antioxidants have been proposed, according to Scheme 4.2, though these mechanisms can occur in parallel.

Mechanism (i) is referred to as H-atom transfer, in which the ArOH itself becomes a radical; hydroxylated rings in chalcones can be easily converted to phenoxy radicals ($ArO^{\cdot}$). Mechanism (ii) is one electron transfer, in which the antioxidant agent gives an electron to the free radical and becomes a radical cation ($ArO^{+\cdot}$).

Experimental results have shown that chalcones have the potential for the scavenging of various radicals, and that the presence of hydroxyl groups may represent a higher efficiency of this activity (Xue et al. 2012). The antioxidant potential of molecules frequently calls attention for their possible therapeutic application in diseases or disorders associated with oxidative stress, such as inflammation and cancer (Padhye et al. 2010; Sikander et al. 2011). However, as described before, oxidative processes can also occur in biodiesel, representing a serious disadvantage for its usage.

In light of the scientific reports and factors associating hydroxychalcones with antioxidant activity, herein we report our study of the influence of two hydroxylated chalcones: (chalcones 16a and 16b; Fig. 4.13) in the oxidation of biodiesel.

### 4.2.1 Synthesis of Chalcones

Chalcones can be found as naturally in plants, fruits, and vegetables, which contain the key enzyme chalcone synthase (CSH). CHS is a member of the plant polyketide synthase superfamily and is structurally classified as a type III polyketide synthase enzyme. As outlined in Fig. 4.14, CHS catalyzes the sequential condensation of three acetate molecules derived from malonyl-CoA and *p*-coumaroyl-CoA (CoA = coenzyme A). CHS together with stilbene synthase (STS) generates the chalconaringenin (2′,4′, 6′,4-tetrahydroxychalcone), which can be further cyclized to narigenin by action of chalcone isomerase (CHI) (step not shown, see Dao et al. 2011 for more details). The action of CHS with chalcone reductase (CHR) can

**Scheme 4.2** Proposed mechanisms for antioxidant activity (Xue et al. 2012)

$$R^{\bullet} + ArOH \longrightarrow RH + ArO^{\bullet} \qquad (i)$$

$$R^{\bullet} + ArOH \longrightarrow R^{\ominus} + ArO^{+\bullet} \qquad (ii)$$

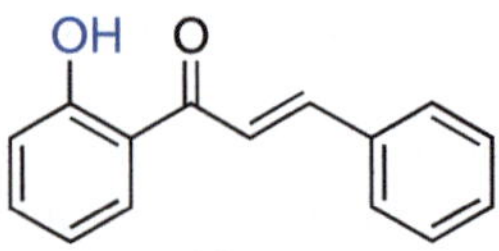

**16a**

(*E*)-1-(2-hydroxyphenyl)-3-phenyl prop-2-en-1-one

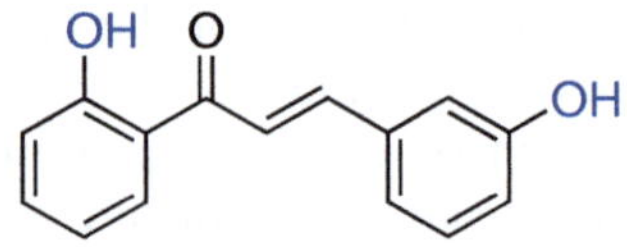

**16b**

(*E*)-1-(2-hydroxyphenyl)-3-(3-hydroxyphenyl) prop-2-en-1-one

**Fig. 4.13** Structures of chalcones 16a and 16b synthesized and assessed for antioxidant activity in biodiesel

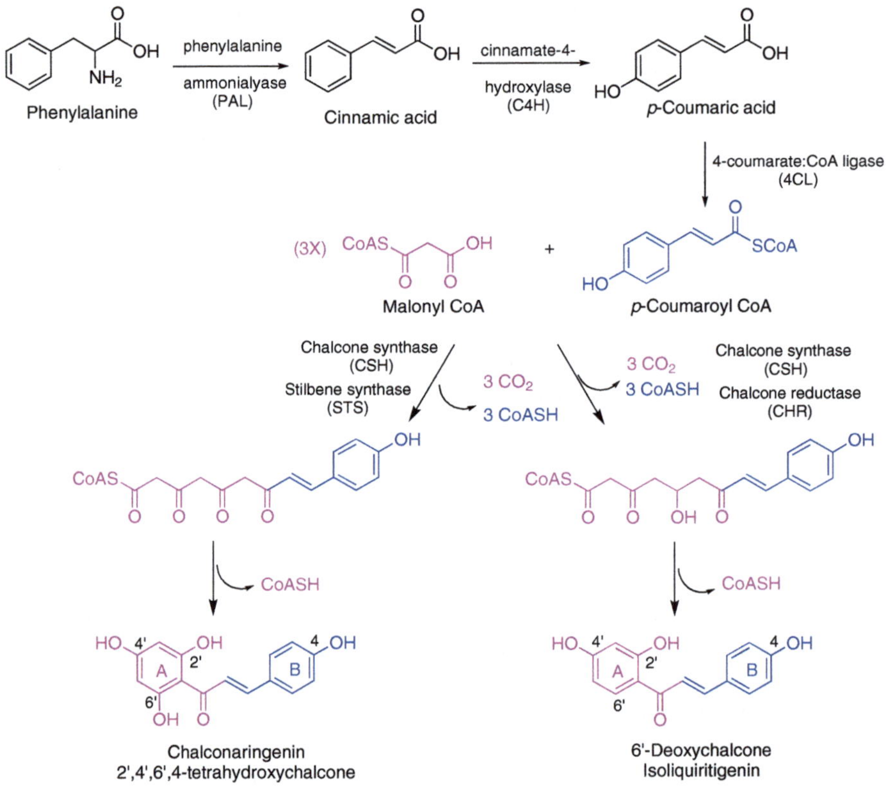


**Fig. 4.14** Overview of the biosynthesis of chalcones: chalconaringenin (tetrahydroxychalcone) and isoliquiritigenin (6′-deoxychalcone). Modified from Veith and Grayer (2006) and Dao et al. (2011)

generate a 6′-deoxychalcone known as isoliquiritigenin (Veith and Grayer 2006; Dao et al. 2011; Orlikova et al. 2011).

Other cyclization reactions are possible besides the one that yields chalcones, and CHS can accept other substrates as starting materials to generate corresponding chalcones and lactone (cyclic esters) products, such as feruloyl-CoA, phenylacetyl-CoA, butyryl-CoA (Dao et al. 2011).

Nowadays, it is very practical to obtain series of chalcones by synthetic methods, and with high purities, which avoids many of the obstacles faced when these products are isolated from natural sources (Detsi et al. 2009; Aly et al. 2014; Mathew et al. 2017). The classical type of reaction carried to obtain chalcones is the Claisen–Schmidt condensation, in which a ketone reacts with an aromatic aldehyde, generally under basic catalysis (Dawane et al. 2009). This is possible because ketones, such as propanone or acetophenone, submitted to basic conditions form enolates, which react with the carbonyl group of aldehydes. It also involves dehydration to yield a product in which the double bond is conjugated both to a carbonyl group and to an aromatic ring as follows (Scheme 4.3).

An acceptable proposed mechanism for the formation of chalcones by Claisen–Schmidt condensation is shown in Fig. 4.15 (Zhuang et al. 2017).

The mechanism of Claisen–Schmidt condensation depicted in Fig. 4.15 shows that, initially, the oxygen of a base catalyst (such as NaOH or KOH) captures an acid hydrogen (α-hydrogen) from acetophenone, forming an enolate, which presents keto–enol tautomerism (shown in brackets). The enolate is a good nucleophile

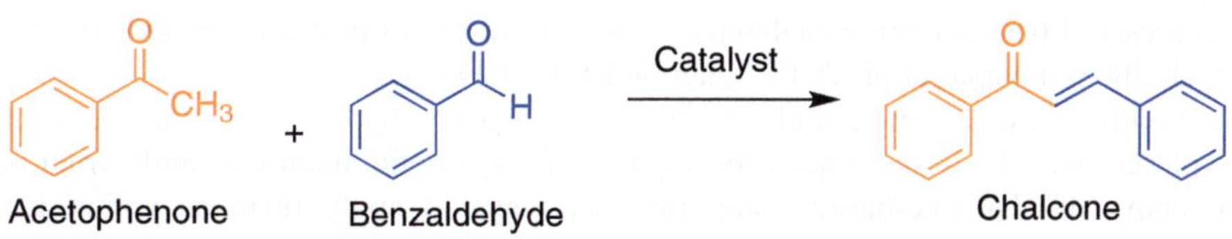


**Scheme 4.3** Claisen–Schmidt condensation reaction (Chavan et al. 2016)

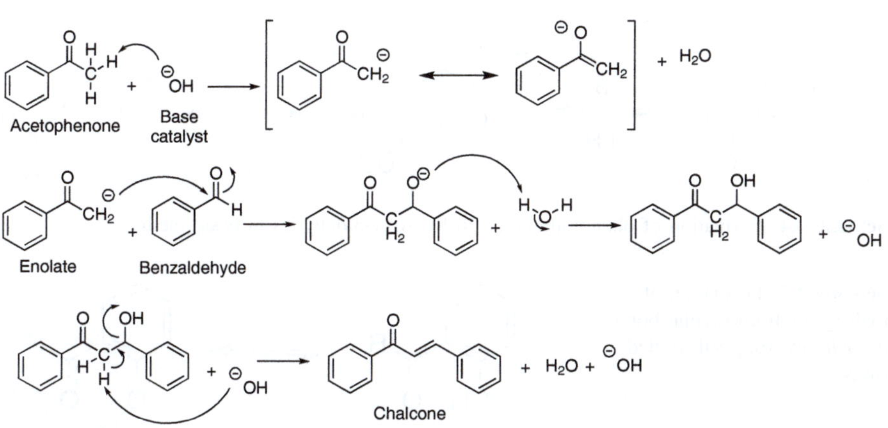


**Fig. 4.15** Mechanism of Claisen–Schmidt condensation reaction to obtain chalcones

and attacks the partially positive carbonyl of the aromatic aldehyde, forming a tetrahedral intermediate (alkoxide). Under basic conditions, the alkoxide intermediate is protonated and the α-hydrogen (from the methylene group) is then removed, favoring the dehydration final step to yield the chalcone and to regenerate the basic media.

The synthesis of chalcones containing hydroxyl group (s) under basic conditions can be problematic because the electrophilicity of the aldehyde carbonyl carbon is reduced. The hydrogen from hydroxyl groups can be extracted by bases, forming a less reactive phenoxide anion (Scheme 4.4) or, in case of hydroxyl groups in the *ortho* position of the ring, intramolecular bonding with the carbonyl oxygen can be formed (Scheme 4.5) (Rozmer and Perjési 2016).

These factors result in decreased or no product formation so, in order to work around this problem, some authors use hydroxyl protective groups under basic conditions, such as methoxymethyl groups (MOM). The protective group is inserted prior to the Claisen–Schmidt condensation, which can be removed after the formation of the hydroxy-chalcone under acidic conditions (Scheme 4.6) (Nishida and Kawabata 2006; Detsi et al. 2009; Zhuang et al. 2017).

An alternative approach is the use of acids as catalysts for the synthesis of chalcones, which proceeds with the formation of an enol. Although both ketone and aldehyde need to be activated in acidic conditions, it can be directly applied for the synthesis of base-sensitive chalcones without prior use of protective groups (Petrov et al. 2008; Jayapal et al. 2010; Qian and Liu 2011).

In this way, the potential of chalcones, mainly hydroxychalcones, can be explored based on the relatively rapid, safe synthetic methods, and structure diversity of chalcone-based compounds in ways to apply them as antioxidant agents.

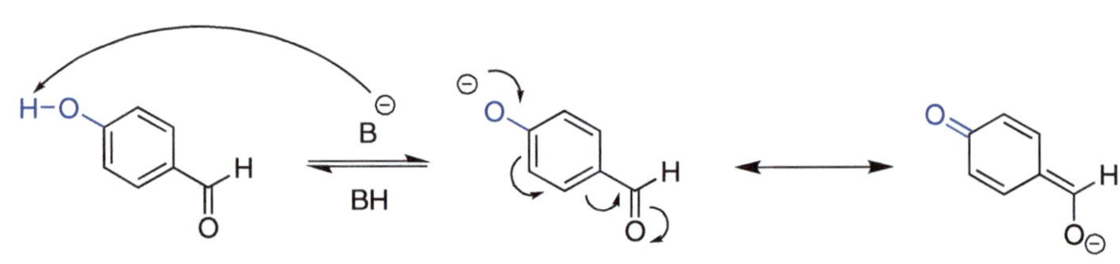


**Scheme 4.4** Formation of phenoxide anion in hydroxy-aldehydes in basic media

**Scheme 4.5** Formation of hydrogen intramolecular bond in *ortho*-hydroxy-substituted rings

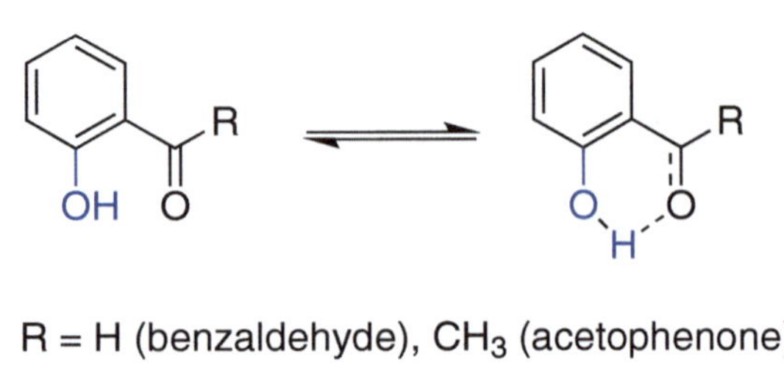

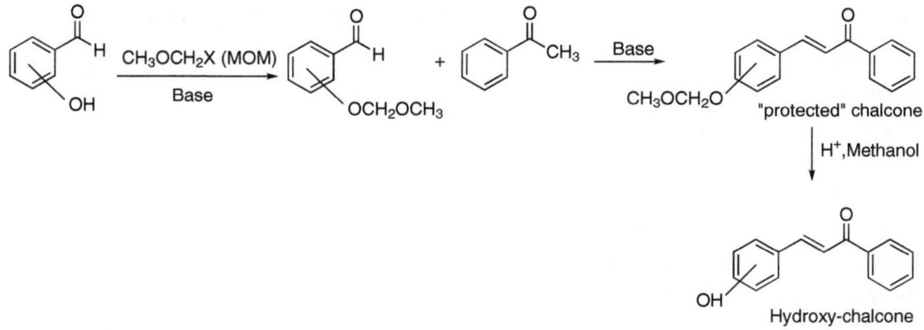


**Scheme 4.6** Synthesis of hydroxyl chalcones using methoxymethyl (MOM) protective group

## 4.3 Methodology

### *4.3.1 Synthesis and Analysis of Biodiesel*

Soybean frying oil was obtained from the restaurant of a University (Federal University of Pelotas, Brazil). Sequentially, the material was mixed with Celite in a proportion of 2 parts of oil to 4 parts of waste cooking oil. Afterward, the purified soybean oil was transesterified according to the methodology proposed by De Oliveira et al. (2013). The reaction was performed by the addition of 40 mL of a solution of sodium hydroxide 2.5% (m/v) in methanol in 100 mL of soybean oil, which was kept under reflux at 70 °C for one hour. The material was cooled to room temperature, then 60 mL of methanol and 1.5 mL of concentrated sulfuric acid were added to the system. The reaction was allowed to continue for another hour under reflux at 70 °C. Finally, the product was washed three times with distilled water, filtered through anhydrous sodium sulfate, and dried in a rotary evaporator.

Kinematic viscosity, acid value, iodine value, and flash point were determined according to ASTM standard methods. Saponification value was determined according to the AOAC method (AOAC 1997). The standard methods utilized to determine biodiesel parameters are shown in Table 4.2.

Fatty acid methyl esters (FAME) obtained from the transesterification reaction were identified in a gas chromatograph equipped with a flame ionization detector (GC-FID) (Shimadzu QP-2010). Quantitative analysis was carried out by area

**Table 4.2** Standard methods for the characterization of biodiesel

| Parameter | Standard method |
|---|---|
| Kinematic viscosity | ASTM D445(ASTM 2006) |
| Acid value | ASTM D664-67 (ASTM 2011) |
| Iodine value | ASTM D459-97 (ASTM 1997) |
| Saponification value | AOCS CD3-25 (AOAC 1997) |
| Flash point | ASTM 93-94 (ASTM 1995) |

**Table 4.3** Chromatographic analysis of fatty acid methyl esters from soybean frying oil

| Parameter | Conditions |
|---|---|
| Column | (100 m × 0.25 m × 0.25 μm) |
| Carrier gas | Nitrogen |
| Injection volume | 1 μL |
| Split ratio | 1:25 |
| Pressure | 44.695 ψ |
| Flow | 1.2 mL $min^{-1}$ |
| Injector temperature | 250 °C |
| Heating ramp | 100 °C (5 min)–7 °C $min^{-1}$ to 200 °C (5 min)–5 °C $min^{-1}$ to 230 °C (10 min) |
| Detector | Flame ionization detector |
| Detector temperature | 250 °C |

normalization using the GC Solution software in comparison to the FAME Mix $C_4C_{24}$ standard (Supelco). Details of chromatographic conditions can be seen in Table 4.3.

### 4.3.2 *Synthesis and Characterization of HydroxyChalcones*

Solvents and chemicals used in the synthesis were obtained from Sigma-Aldrich Co., St. Louis, MO, USA and used without further purification. Chalcones (*E*)-1-(2-hydroxyphenyl)-3-phenylprop-2-en-1-one (16a) and (*E*)-1-(2-hydroxyphenyl)-3-(3-hydroxyphenyl) prop-2-en-1-one (16b) were synthesized by Claisen–Schmidt condensation according to Wang et al. (2014) (Scheme 4.7). A solution of sodium hydroxide 40% in water (2.0 mL) was added to a round bottom flask containing 15 mL of ethanol and 1.0 mmoL of 2-hydroxyacetophenone, in an ice bath. After 30 min, 1.2 mmoL of the substituted aldehyde was added and the mixture was stirred for 12 h, at room temperature. The reaction was monitored by thin-layer chromatography in hexane/ethyl acetate 70/30%. The reaction mixture was neutralized with hydrochloric acid, and the filtered precipitate was purified by recrystallization with ethanol. Melting points were determined using capillary tubes on a Fisatom 430 apparatus. The synthetic chalcones had their structure and purity confirmed by gas chromatography coupled to mass spectrometry (GC-MS) equipped with an AOC-20i automatic injector and a Rtx-5MS 30 m × 0.25 mm × 0.25 μm column.

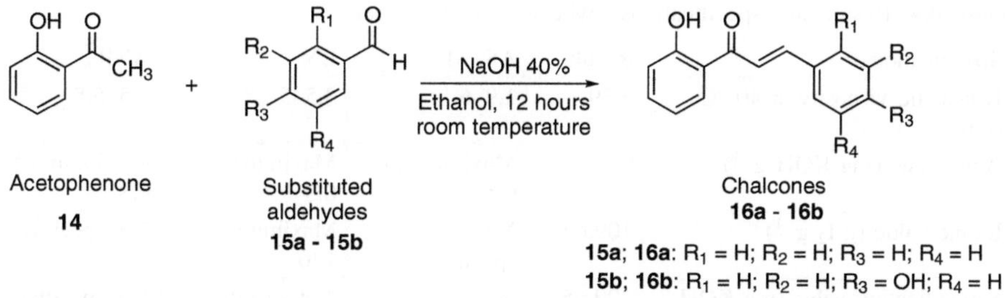


**Scheme 4.7** Claisen–Schmidt condensation reaction to obtain chalcones 16a and 16b

### *4.3.3 Thermal Analysis*

Oxidative stability of biodiesel without and with the addition of hydroxychalcones was assessed by differential scanning calorimetry (DSC), which was carried on a 200 F3 Maia equipment model (NETZSCH-GERÄTEBAU GMBH). Samples of biodiesel and chalcones were analyzed in closed aluminum pans, in triplicate, separately or mixed, in which the final concentration of chalcone in biodiesel was 1250 ppm. The experiments were conducted under nitrogen atmosphere at a flow rate of 50 mL min$^{-1}$ and under constant temperature heating of 10 °C min$^{-1}$ from 20 to 500 °C.

## 4.4 Results and Discussion

The crude waste cooking oil used in this work had a dark yellow and a viscous appearance due to its constant use, which is also linked to its high acidity and unpleasant odor. In order to be used as a fuel, it needed to surpass a pretreatment process with Celite so that most of its impurities could be eliminated. The clean up process was very efficient to remove most of the contaminants in the crude oil, such as solid particles and water, turning its color into a light yellow. These substances could interfere in the transesterification reaction leading to poor yields and a biodiesel with low quality (César et al. 2017).

The results of the physicochemical analysis of the biodiesel obtained from fried soybean oil (FSB) are shown in Table 4.4. These results can be compared to the biodiesel quality norms established in the USA (ASTM), European Union (EN), and Brazil (ANP).

According to Table 4.4, soybean frying biodiesel met all the regulated physicochemical parameters tested. Between tests was iodine value, which measures the degree of unsaturation of the fatty esters from the sample. The saponification value measures the amount of alkali needed to saponify a certain fraction of an oil or fat and usually increases as the length of the carbon chain decreases. Thus, it serves as a tool to indicate the formation of low molecular weight compounds throughout the

**Table 4.4** Results and specifications soybean frying biodiesel

| Specification | Result | ASTM | EN | ANP |
|---|---|---|---|---|
| Kinematic viscosity at 40 °C ($mm^2$ $s^{-1}$) | 4.39 | 1.9–6 | 3.5–5 | 2,5–5,5 |
| Acid value (mg KOH $g^{-1}$) | 0.5 | Maximum of 0.5 | Maximum of 0.5 | Maximum of 0.8 |
| Iodine value (g $I_2$ $g^{-1}$) | 109.65 | Not specified | Maximum of 120 | Not specified |
| Saponification value (mg KOH $g^{-1}$) | 200.65 | Not specified | Not specified | Not specified |
| Flash point (°C) | 165 | Minimum of 93 | Minimum of 101 | Minimum of 101 |

oxidation of biodiesel. Since this property is only a relative measure to determine the length of the carbon chains in oil, US and European standards do not regulate it (Knothe and Razon 2017).

In turn, the acid value is the amount of alkali needed to neutralize a certain fraction of an oil or fat can be used to determine the presence of free fatty acids in the biodiesel sample. Therefore, it is an important parameter to regulate the quality of biodiesel since high acid values mean parts of the engine can be corroded (Georgogianni et al. 2007). Consecutively, kinematic viscosity is the amount of time needed for a liquid at 40 °C to flow through a viscometer. Generally, the viscosity of a liquid decreases as its level of unsaturation increases. Parallel to density, it also increases as the chain length of fatty acids becomes bigger or in the presence of contaminants such as soaps. This property is related to the amount of combustible injected in the engine and, thus, is of extreme relevance to the quality of biodiesel (Encinar et al. 2007).

Chromatographic analysis (Table 4.5; Fig. 4.16) revealed that soybean biodiesel is composed of several fatty acids. Nonetheless, linoleic acid (C18:2*n*6c) played a dominant role in the lipidic profile representing more than half (51.79%) of the composition of the biodiesel. Oleic acid (C18:1*n*9c) and palmitic acid (C16:0) were also important constituents representing 28.75 and 11.13%. Other biomolecules such as linolenic (C18:3*n*3) and stearic acid (C18:0) were found in small quantities. Since more than 85% of the fatty acids in the biodiesel were unsaturated, it can be said that the oxidative stability was relatively lower because these acids are more reactive toward oxidation reactions (Sajjadi et al. 2016).

In general, the transesterification reaction presented yields close to 96%. In a similar work, Georgogianni et al. (2007) isolated methyl esters of soybean frying oil in yields varying from 42 to 97% under mechanical stirring. Encinar et al. (2007) performed ethanolysis in waste frying oil under alkaline catalysis with yields ranging from 46.8 to 74.2% of oil conversion into esters.

**Table 4.5** Fatty acid composition of soybean biodiesel

| Fatty acid | Percentage (per area) | Peak number in Fig. 4.16 |
|---|---|---|
| C16:0 | 11.13 | 1 |
| C18:0 | 3.85 | 2 |
| C18:1*n*9c | 28.75 | 3 |
| C18:2*n*6c | 51.79 | 4 |
| C18:3*n*3 | 4.46 | 5 |
| ∑saturated | 14.98 | |
| ∑unsaturated | 85.02 | |

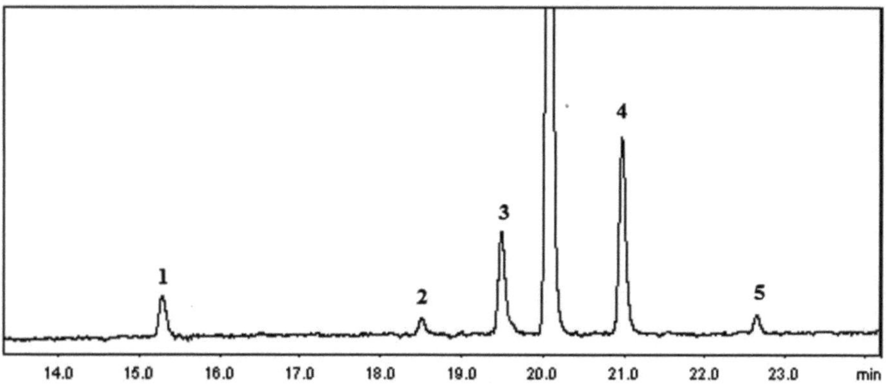


**Fig. 4.16** Chromatogram of fatty acids in waste soybean cooking oil

In relation to the synthetic chalcones, two products were successfully obtained in moderate yields of 46% for chalcone 16a and 32% for chalcone 16b. Wang et al. (2014) obtained the exact structure of chalcone 16a in a 55.7% yield. Kamboj et al. (2010) obtained the exact structure of chalcone 16b in a 52% yield under microwave irradiation, and Ohkatsu and Satoh (2008) obtained this product in a 13.9% yield under conventional conditions (without use of microwaves). Data of chalcones are shown in Table 4.6.

Results from spectrometric analysis are shown below. The chromatogram and mass spectrum of chalcone 16a are shown in Figs. 4.17 and 4.18, and chromatogram and mass spectrum for chalcone 16b are shown in Figs. 4.19 and 4.20.

It can be seen from the physical properties determined experimentally that the synthesized chalcones match the expected parameters, so their structures could be confirmed. Chalcone 16a presented an experimental melting point of 90–91 °C, which was very proximate to the result found in the literature for the same synthetic compound (89 °C) (Jhala et al. 2006). The same could be observed for chalcone 16b, which had an experimental melting point of 163–165 °C, proximate to the one found by Kamboj et al. (2010) (165–167 °C). It is common to observe some variations in experimental melting points to the ones found by other researchers,

**Table 4.6** Data of synthesized hydroxychalcones

| Code | Structure | Melting point (°C) | Exact mass |
|---|---|---|---|
| Chalcone 16a (*E*)-1-(2-hydroxyphenyl)-3-phenylprop-2-en-1-one | 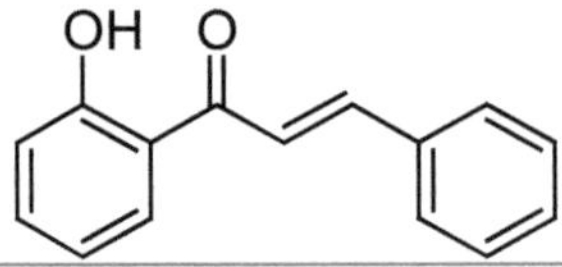 | Exp.: 90–91 Lit.: 89 (Jhala et al. 2006) | 224.08 |
| Chalcone 16b (*E*)-1-(2-hydroxyphenyl)-3-(3-hydroxyphenyl) prop-2-en-1-one | 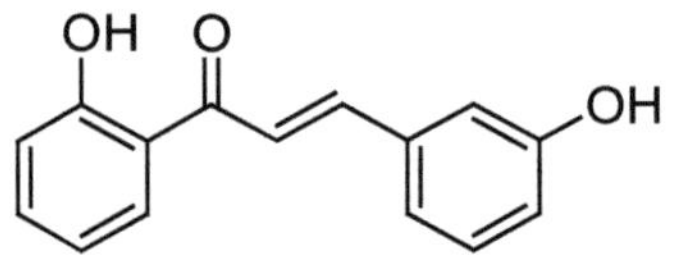 | Exp.: 163–165 Lit.: 165–167 (Kamboj et al. 2010) | 240.08 |

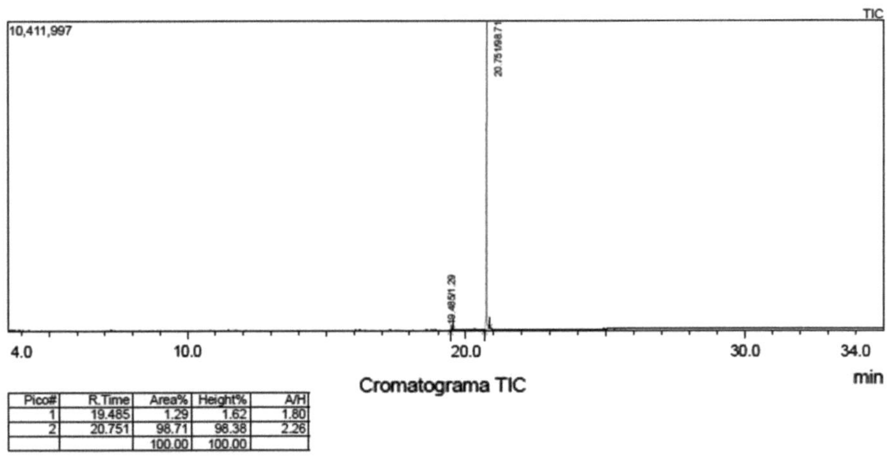


**Fig. 4.17** Chromatogram of chalcone 16a (*E*)-1-(2-hydroxyphenyl)-3-phenylprop-2-en-1-one

mainly due to the different manipulations or solvents used in the recrystallization procedures. Chromatograms of both synthetic products show peaks at retention times of around 20 min, and the integration of the peaks generated the mass spectrum, which were in accordance to the exact masses of the chalcones. Figure 4.19 presents a molecular ion of m/z 224 in accordance to the exact mass of chalcone 16a (224.08), and Fig. 4.21 presented a molecular ion of 240 respective to the mass of 240.08 of chalcone 16b.

In the last few years, thermal analysis by DSC has been commonly used for the purpose of antioxidant activity evaluation (Dunn 2008). Two major quality control tests are assessed by DSC in industry: tests conducted under isothermal conditions indicate oxidation induction time (OIT), and the ones conducted at different temperature scans indicate the oxidation onset temperature (OOT). The difference

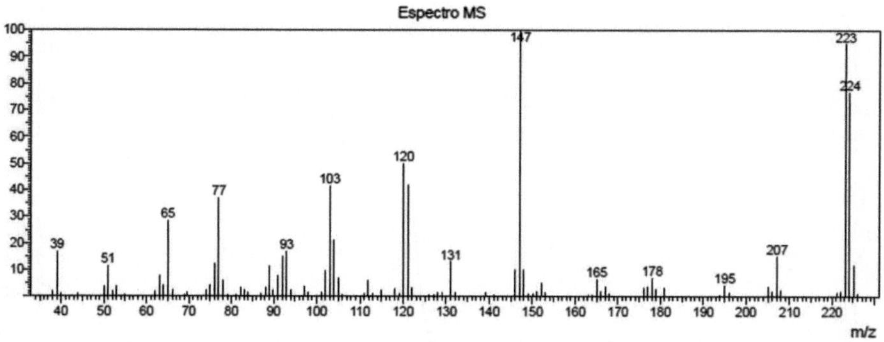


**Fig. 4.18** Mass spectrum of chalcone 16a (*E*)-1-(2-hydroxyphenyl)-3-phenylprop-2-en-1-one

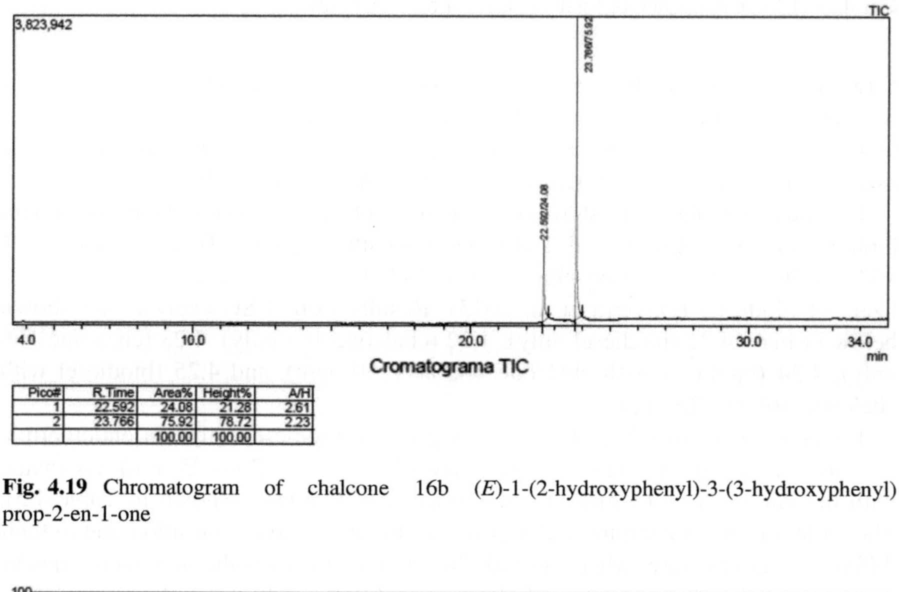


| Pico# | R.Time | Area% | Height% | A/H |
|---|---|---|---|---|
| 1 | 22.592 | 24.08 | 21.28 | 2.61 |
| 2 | 23.766 | 75.92 | 78.72 | 2.23 |
| | | 100.00 | 100.00 | |

**Fig. 4.19** Chromatogram of chalcone 16b (*E*)-1-(2-hydroxyphenyl)-3-(3-hydroxyphenyl) prop-2-en-1-one

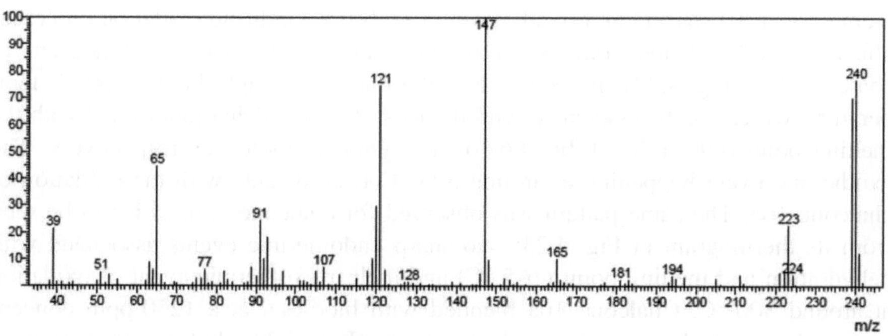


**Fig. 4.20** Mass spectrum of chalcone 16b (*E*)-1-(2-hydroxyphenyl)-3-(3-hydroxyphenyl) prop-2-en-1-one

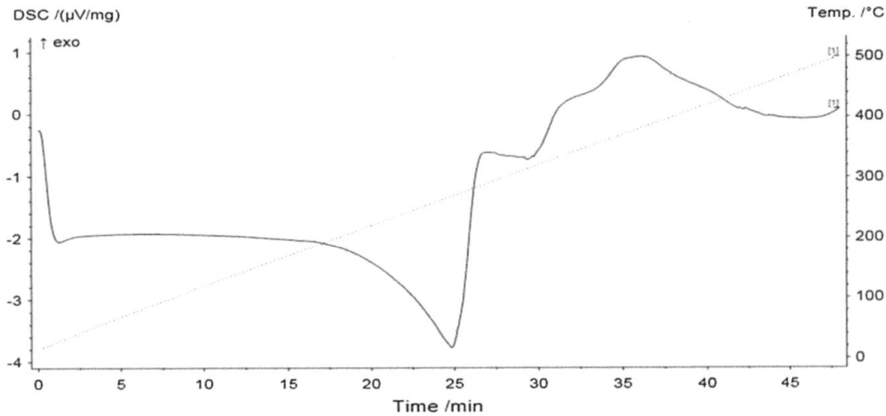


**Fig. 4.21** DSC thermogram of fried soybean oil biodiesel only

between these calorimetric tests is that, under isothermal conditions, oxidation is registered as the time the sample exhibits a sudden exothermic event, whereas under temperature scanning experiments, oxidation is registered as the temperature the exothermic event occurs (Focke and Van Der Westhuizen 2010).

Recently, our laboratory showed the green synthesis of a series of chalcones with biological activity (Ritter et al. 2015) and deposited a patent (Brazilian patent BR 1020130300497) of hydroxy-functionalized chalcones applied as oxidative stabilizers in biodiesel (Oliveira et al. 2013). Results from DSC analysis are shown below in Figs. 4.21 (biodiesel only), 4.22 (chalcone 16a only), 4.23 (chalcone 16b only), 4.24 (biodiesel with chalcone 16a at 1250 ppm), and 4.25 (biodiesel with chalcone 16b at 1250 ppm).

It can be seen from Fig. 4.21 (thermogram of biodiesel only) an endothermic (downwards) event at a temperature range of 200–270 °C (~25 min) associated with the vaporization of methyl esters present in biodiesel (shown in Table 4.5). The variation of temperature will depend on the fatty acid composition due to their differences in structure (Misutsu et al. 2015). The other exothermic events (peaks facing upwards) starting at around 30 min (~300 °C) indicate oxidation processes affecting biodiesel and combustion of unsaturated fatty acids. Thermogram of chalcone 16a (Fig. 4.22) shows two endothermic events right before 100 °C temperature, which can be associated with the loss of water (dehydration) and with the melting point (89–91 °C; Table 4.6) of the synthetic molecule, respectively. The exothermic event happening at around 325 °C is associated with the oxidation of chalcone 16a. The same pattern was observed for chalcone 16b, as it can be seen from its thermogram in Fig. 4.23: two sharp endothermic events associated with dehydration and melting point (165 °C) and a clear exothermic event of oxidation at around 300 °C. Chalcone 16a blended with biodiesel at a 1250 ppm concentration does not show an antioxidant activity (Fig. 4.24), but rather caused an increase of the temperature in which biodiesel started to suffer oxidation, or an

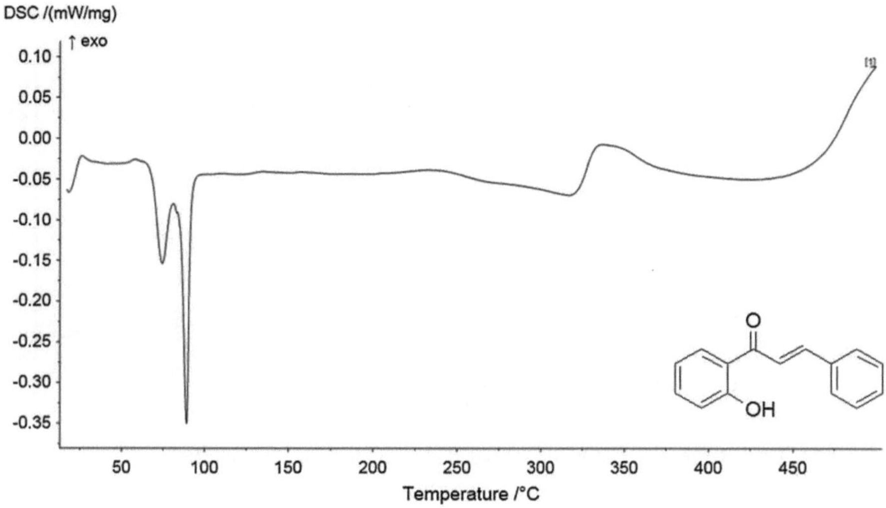


**Fig. 4.22** DSC thermogram of chalcone 16a only

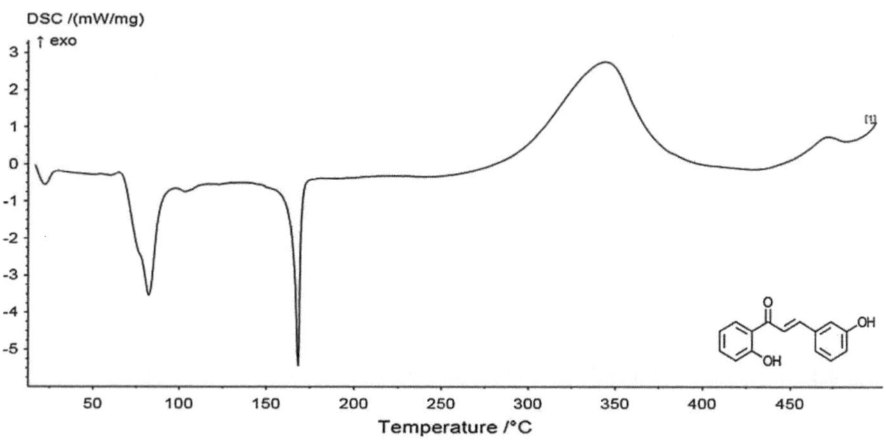


**Fig. 4.23** DSC thermogram of chalcone 16b only

increase in OOT. Chalcone 16b showed a potent antioxidant effect as the thermogram depicted in Fig. 4.25 lacks any exothermic events of oxidation. It seems that events happening with the chalcone overlap the one observed in biodiesel (~25 min), and that the chalcone could be suffering or neutralizing oxidation events itself instead of biodiesel. The other slightly distorted endothermic event (27–35 min in Fig. 4.25) could be due to the formation of other stable products resulting from the oxidation of chalcone 16b. Their results showed that the chalcone 16b containing two hydroxyl groups was more effective than chalcone 16a, which has only one, but that both synthetic molecules influenced in the oxidative stability

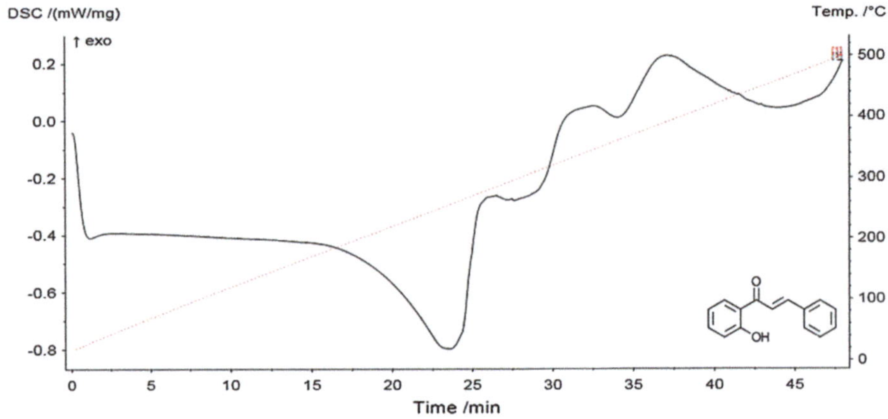


**Fig. 4.24** DSC thermogram of biodiesel blended with chalcone 16a at 1250 ppm

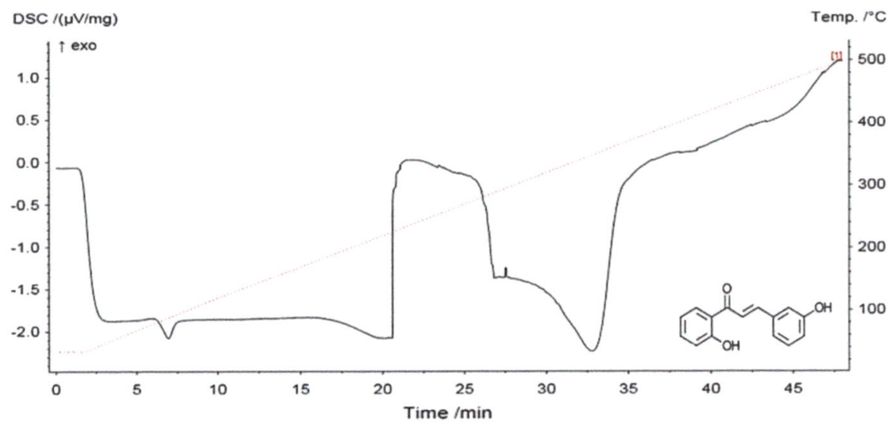


**Fig. 4.25** DSC thermogram of biodiesel blended with chalcone 16b at 1250 ppm

of biodiesel. It is important to note the sample of biodiesel in this study was obtained from used frying oil, which itself has already been exposed to oxidative and hydrolysis processes. Also, the 1250 ppm concentration of chalcones blended with biodiesel is within the range of 500–2000 ppm normally used in similar studies (Dunn 2008). It is likely that a higher concentration of these compounds mixed with biodiesel would induct a more notable antioxidant effect. Thus, synthetic molecules based on the structural features discussed here are still promising agents to enhance biodiesel oxidative stability.

## 4.5 Conclusion

Biodiesel has many advantages compared to the ephemeral petrodiesel, but its potential has not yet been fully explored due to some drawbacks. In this chapter, it was demonstrated that oxidative stability is one of the major limitations associated with biodiesel. In this way, it is very important to search for agents that can avoid oxidative processes or enhance biodiesel stability. Phenolic compounds, specially natural or synthetic chalcones containing hydroxyl groups, have shown interesting antioxidant effects and, therefore, should be further explored as antioxidant agents or stabilizers to be applied along with biodiesel production. Finally, calorimetric analysis performed by DSC proved to be a very helpful tool to assess potential antioxidant effects resulting from the addition of isolated or synthetic compounds in biodiesel.

## References

Aly MRES, Fodah HHAER, Saleh SY (2014) Antiobesity, antioxidant and cytotoxicity activities of newly synthesized chalcone derivatives and their metal complexes. Eur J Med Chem 76:517–530

Association of Official Analytical Chemists (AOAC) (1997) Section C: Commercial Fats and Oils, American Oil Chemists Society (AOCS) official method Cd 3–25 for saponification value. Association of Official Analytical Chemists, Washington

ASTM D (1997) Standard test method for iodine value of drying oils and fatty acids. Int Annual book of ASTM standards

ASTM Standard (1995) D93: standard test methods for flash point by Pensky-Martens closed cup tester. American Society for Testing and Materials, Philadelphia, PA (USA)

ASTM Standard (2006) D445: standard test method for kinematic viscosity of transparent and opaque liquids (and calculation of dynamic viscosity). American Society for Testing and Materials, West Conshohocken, PA (USA)

ASTM Standard (2011) Standard test method for acid number of petroleum products by potentiometric titration-ASTM D 664-11A. ASTM International, West Conshohocken, PA

Atabani AE, Silitonga AS, Ong HC et al (2013) Non-edible vegetable oils: a critical evaluation of oil extraction, fatty acid compositions, biodiesel production, characteristics, engine performance and emissions production. Renew Sustain Energy Rev 18:211–245

Cao G, Sofic E, Prior RL (1997) Antioxidant and prooxidant behavior of flavonoids: structure-activity relationships. Free Radic Biol Med 22:749–760

Chavan BB, Gadekar AS, Mehta PP, et al (2016) Synthesis and medicinal significance of chalcones—a review. Asian J Biomed Pharm Sci 56:1–7

da César AS, Werderits DE, de Oliveira Saraiva GL, da Guabiroba RCS (2017) The potential of waste cooking oil as supply for the Brazilian biodiesel chain. Renew. Sustain. Energy Rev 72:246–253

Dantas MB, Albuquerque AR, Barros AK et al (2011) Evaluation of the oxidative stability of corn biodiesel. Fuel 90:773–778

Dao TTH, Linthorst HJM, Verpoorte R (2011) Chalcone synthase and its functions in plant resistance. Phytochem Rev 10:397–412

Dawane BS, Konda SG, Shaikh BM, Bhosale RB (2009) An improved procedure for synthesis of some new 1,3-diaryl-2-propen-1-ones using PEG-400 as a recyclable solvent and their antimicrobial evaluation. Acta Pharm 59:473–482

De Oliveira DM, Ongaratto DP, Fontoura LAM et al (2013) Obtenção de biodiesel por transesterificação em dois estágios e sua caracterização por cromatografia gasosa óleos e gorduras em laboratório de química orgânica. Quim Nova 36:734–737

Detsi A, Majdalani M, Kontogiorgis CA et al (2009) Natural and synthetic 2′-hydroxy-chalcones and aurones: synthesis, characterization and evaluation of the antioxidant and soybean lipoxygenase inhibitory activity. Bioorganic Med Chem 17:8073–8085

Días-Tielas C, Graña E, Reigosa MJ, Sánchez-Moreiras AM (2016) Biological activities and novel applications of chalcones. Planta Daninha 34:607–616

Dunn RO (2006) Oxidative stability of biodiesel by dynamic mode pressurized–differential Scanning Calorimetry (P–Dsc). Am Soc Agric Biol Eng 49:1633–1641

Dunn RO (2008) Antioxidants for improving storage stability of biodiesel. Biofuels, Bioprod Biorefining 2:304–318

Encinar JM, González JF, Rodríguez-Reinares A (2007) Ethanolysis of used frying oil. Biodiesel preparation and characterization. Fuel Process Technol 88:513–522

Focke WW, Van Der Westhuizen I (2010) Oxidation induction time and oxidation onset temperature of polyethylene in air: testing Gimzewski's postulate. J Therm Anal Calorim 99:285–293

Franceschelli S, Pesce M, Vinciguerra I et al (2011) Licocalchone-C extracted from glycyrrhiza glabra inhibits lipopolysaccharide-interferon-gamma inflammation by improving antioxidant conditions and regulating inducible nitric oxide synthase expression. Molecules 16:5720–5734

Fu J, Hue BTB, Turn SQ (2017) Oxidation stability of biodiesel derived from waste catfish oil. Fuel 202:455–463

Georgogianni KG, Kontominas MG, Tegou E, et al (2007) Biodiesel production: reaction and process parameters of alkali-catalyzed transesterification of waste frying oils biodiesel production: reaction and process parameters of alkali-catalyzed transesterification of waste frying oils. Energy 3023–3027. https://doi.org/10.1021/ef070102b

Gui MM, Lee KT, Bhatia S (2008) Feasibility of edible oil versus non-edible oil versus waste edible oil as biodiesel feedstock. Energy 33:1646–1653

Heim KE, Tagliaferro AR, Bobilya DJ (2002) Flavonoid antioxidants: chemistry, metabolism and structure-activity relationships. J Nutr Biochem 13:572–584

Herbinet O, Pitz WJ, Westbrook CK (2010) Detailed chemical kinetic mechanism for the oxidation of biodiesel fuels blend surrogate. Combust Flame 157:893–908

Hoekman SK, Broch A, Robbins C et al (2012) Review of biodiesel composition, properties, and specifications. Renew Sustain Energy Rev 16:143–169

Jayapal MR, Sreenivasa Prasad K, Sreedhar NY (2010) Synthesis and characterization of 2,4-dihydroxy substituted chalcones using aldol condensation by $SOCl_2$/EtOH. J Chem Pharm Res 2:127–132

Jhala YS, Dulawat SS, Verma BL (2006) Solvent-free improved syntheses of some substituted 1, 3-diaryl-propenones and 3, 5-diaryl-6-carbethoxycyclohexenones under microwave irradiation and their antibacterial activity. Indian J Chem 45:466–469

Kamboj RC, Arora R, Sharma G et al (2010) Eco-friendly synthesis and antimicrobial activity of chalcones. Der Pharma Chem 2:157–170

Knothe G (2005) Dependence of biodiesel fuel properties on the structure of fatty acid alkyl esters. Fuel Process Technol 86:1059–1070

Knothe G (2006) Analyzing biodiesel: standards and other methods. JAOCS J Am Oil Chem Soc 83:823–833

Knothe G (2007) Some aspects of biodiesel oxidative stability. Fuel Process Technol 88:669–677

Knothe G, Razon LF (2017) Biodiesel fuels. Prog Energy Combust Sci 58:36–59

Kulkarni MG, Dalai AK (2006) Waste cooking oil - An economical source for biodiesel: a review. Ind Eng Chem Res 45:2901–2913

Kumar N (2017) Oxidative stability of biodiesel: causes, effects and prevention. Fuel 190:328–350

Lapuerta M, Herreros JM, Lyons LL et al (2008) Effect of the alcohol type used in the production of waste cooking oil biodiesel on diesel performance and emissions. Fuel 87:3161–3169

Lee CK, Son SH, Park KK et al (2008) Licochalcone a inhibits the growth of colon carcinoma and attenuates cisplatin-induced toxicity without a loss of chemotherapeutic efficacy in mice. Basic Clin Pharmacol Toxicol 103:48–54

Marcon NS, Colet R, Balen DS et al (2017) Enzymatic biodiesel production from microalgae biomass using propane as pressurized fluid. Can J Chem Eng 95:1340–1344

Mathew B, Adeniyi AA, Joy M et al (2017) Anti-oxidant behavior of functionalized chalcone-a combined quantum chemical and crystallographic structural investigation. J Mol Struct 1146:301–308

Mathiesena L, Malterud KE, Sund RB (1996) Hydrogen bond formation as basis for radical scavenging activity: a structure-activity study of C-methylated dihydrochalcones from Myrica gale and structurally related acetophenones. Free Radic Biol Med 22:307–311

Mekhilef S, Siga S, Saidur R (2011) A review on palm oil biodiesel as a source of renewable fuel. Renew Sustain Energy Rev 15:1937–1949

Misutsu MY, Cavalheiro LF, Ricci TG et al (2015) Thermoanalytical methods in verifying the quality of biodiesel. In: Biernat K (ed) Biofuels—status and perspective. InTech, Croatia, pp 251–269

Mittelbach M, Schober S (2003) The influence of antioxidants on the oxidation stability of biodiesel. JAOCS, J Am Oil Chem Soc 80:817–823

Nishida J, Kawabata J (2006) DPPH radical scavenging reaction of hydroxy- and methoxychalcones. Biosci Biotechnol Biochem 70:193–202

Niu H, Wang W, Li J et al (2017) A novel structural class of coumarin-chalcone fibrates as PPARalpha/gamma agonists with potent antioxidant activities: Design, synthesis, biological evaluation and molecular docking studies. Eur J Med Chem 138:212–220

Ohkatsu Y, Satoh T (2008) Antioxidant and photo-antioxidant activities of chalcone derivatives. J Japan Pet Inst 51:298–308. https://doi.org/10.1627/jpi.51.298

Oliveira M, Rockembach CT, Dias D, et al (2013). Aplicação de Chalconas como Reagente Antioxidante em Biodiesel. Brazilian Patent BR 10 2013 030049 7 A2, 22 Nov 2013

Orlikova B, Tasdemir D, Golais F et al (2011) Dietary chalcones with chemopreventive and chemotherapeutic potential. Genes Nutr 6:125–147

Padhye S, Ahmad A, Oswal N et al (2010) Fluorinated 2′-hydroxychalcones as garcinol analogs with enhanced antioxidant and anticancer activities. Bioorganic Med Chem Lett 20:5818–5821

Pereira CMP, Hobuss CB, Maciel JV et al (2012) Biodiesel renovável derivado de microalgas: avanços e perspectivas tecnológicas. Quim Nova 35:2013–2018

Petrov O, Ivanova Y, Gerova M (2008) $SOCl_2$/EtOH: catalytic system for synthesis of chalcones. Catal Commun 9:315–316

Pinto FV (2017) Boletim Mensal do biodiesel. In: Agência Nac. do Petróleo Gás Nat. e Biocombustíveis—ANP. Web Page. http://www.anp.gov.br/wwwanp/publicacoes/boletins-anp/2386-boletim-mensal-do-biodiesel. Accessed 28 Aug 2017

Pinzi S, Garcia IL, Lopez-Gimenez FJ et al (2009) The ideal vegetable oil-based biodiesel composition: a review of social, economical and technical implications. Energy Fuels 23:2325–2341

Pullen J, Saeed K (2012) An overview of biodiesel oxidation stability. Renew Sustain Energy Rev 16:5924–5950

Qian H, Liu D (2011) Synthesis of chalcones via claisen-schmidt reaction catalyzed by sulfonic acid-functional ionic liquids. Ind Eng Chem Res 50:1146–1149

Rezk BM, Haenen GRMM, Van der Vijgh WJF, Bast A (2002) The antioxidant activity of phloretin: The disclosure of a new antioxidant pharmacophore in flavonoids. Biochem Biophys Res Commun 295:9–13

Righi AA, Alves TR, Negri G et al (2011) Brazilian red propolis: unreported substances, antioxidant and antimicrobial activities. J Sci Food Agric 91:2363–2370

Ritter M, Martins RM, Rosa SA et al (2015) Green synthesis of chalcones and microbiological evaluation. J Braz Chem Soc 26:1201–1210

Rizwanul Fattah IM, Hassan MH, Kalam MA et al (2014a) Synthetic phenolic antioxidants to biodiesel: path toward NOx reduction of an unmodified indirect injection diesel engine. J Clean Prod 79:82–90

Rizwanul Fattah IM, Masjuki HH, Kalam MA et al (2014b) Effect of antioxidants on oxidation stability of biodiesel derived from vegetable and animal based feedstocks. Renew Sustain Energy Rev 30:356–370. https://doi.org/10.1016/j.rser.2013.10.026

Rockembach CT, Dias D, Vieira BM et al (2014) Synthesis of biodiesel from grape seed oil using ultrasound irradiation. Rev Virtual Química 6:884–897

Rozmer Z, Perjési P (2016) Naturally occurring chalcones and their biological activities. Phytochem Rev 15:87–120

Sajjadi B, Raman AAA, Arandiyan H (2016) A comprehensive review on properties of edible and non-edible vegetable oil-based biodiesel: composition, specifications and prediction models. Renew Sustain Energy Rev 63:62–92

Saluja RK, Kumar V, Sham R (2016) Stability of biodiesel—a review. Renew Sustain Energy Rev 62:166–181

Sarin R, Sharma M, Sinharay S, Malhotra RK (2007) Jatropha-Palm biodiesel blends: an optimum mix for Asia. Fuel 86:1365–1371

Senevirathne M, Kim S-H, Siriwardhana N et al (2006) Antioxidant potential of Ecklonia cava on reactive oxygen species scavenging, metal chelating, reducing power and lipid peroxidation inhibition. Food Sci Technol Int 12:27–38

Serifi O, Tsopelas F, Kypreou AM et al (2013) Antioxidant behaviour of 2′-hydroxy-chalcones: a study of their electrochemical properties. J Phys Org Chem 26:226–231

Serqueira DS, Fernandes DM, Cunha RR et al (2014) Influence of blending soybean, sunflower, colza, corn, cottonseed, and residual cooking oil methyl biodiesels on the oxidation stability. Fuel 118:16–20

Shahidi F, Wanasundara PK (1992) Phenolic antioxidants. Crit Rev Food Sci Nutr 32:67–103

Sikander M, Malik S, Yadav D et al (2011) Cytoprotective activity of a trans-chalcone against hydrogen peroxide induced toxicity in hepatocellular carcinoma (HepG2) cells. Asian Pacific J Cancer Prev 12:2513–2516

Singh A, Nigam PS, Murphy JD (2011) Renewable fuels from algae: An answer to debatable land based fuels. Bioresour Technol 102:10–16

Singh SP, Singh D (2010) Biodiesel production through the use of different sources and characterization of oils and their esters as the substitute of diesel: a review. Renew Sustain Energy Rev 14:200–216

Stuart BH (2012) Forensic analytical techniques. Wiley, New Jersey, 234 p

Tomás-Barberán FA, Clifford MN (2000) Flavanones, chalcones and dihydrochalcones—nature, occurrence and dietary burden. J Sci Food Agric 80:1073–1080

Veith NC, Grayer RJ (2006) Chalcones, Dihydrochalcones, and Aurones. In: Andersen OM, Markham KR (eds) Flavonoids: chemistry, biochemistry, and applications, pp 1003–1071. CRC Press Taylor & Francis Group, Florida

Vieira BM, Elicker C, Nunes CFP et al (2016) The synthesis and characterization of Butia capitata seed oil as a FAME feedstock. Fuel 184:533–535

Wang F-W, Wang S-Q, Zhao B-X, Miao J-Y (2014) Discovery of 2′-hydroxychalcones as autophagy inducer in A549 lung cancer cells. Org Biomol Chem 12:3062–3070

Wang Y, Pengzhan Liu SO, Zhang Z (2007) Preparation of biodiesel from waste cooking oil via two-step catalyzed process. Energy Convers Manag 48:184–188

Xin J, Imahara H, Saka S (2009) Kinetics on the oxidation of biodiesel stabilized with antioxidant. Fuel 88:282–286

Xue Y, Zheng Y, An L et al (2012) A theoretical study of the structure-radical scavenging activity of hydroxychalcones. Comput Theor Chem 982:74–83

Yaakob Z, Narayanan BN, Padikkaparambil S et al (2014) A review on the oxidation stability of biodiesel. Renew Sustain Energy Rev 35:136–153

Zhuang C, Zhang W, Sheng C et al (2017) Chalcone: a privileged structure in medicinal chemistry. Chem Rev 117:7762–7810

# Chapter 5
# Fluorescence Spectroscopy as an Alternative Analytical Tool for Monitoring Biodiesel Oxidative Stability: Thermal Oxidation Effect on the Endogenous Chromophores and Fluorophores in Biodiesel

**Keurison Figueredo Magalhães, Anderson Rodrigues Lima Caires, Tiago Andrade Chimenez, Mariele Cappelari Fripp, Fabíola Machado and Samuel Leite Oliveira**

**Abstract** Thermal stability of biodiesel is an important quality factor and it must be precisely evaluated. Despite the existence of conventional methods, fast, accurate, and robust analytical procedures are needed and still being developed. In this study, soybean and canola biodiesels underwent degradation by heating the samples and a systematic dependence on the concentration of endogenous chromophores (conjugated dienes, trienes, and tetraenes) and fluorophores (conjugated tetraenes) was optically monitored. UV–Vis absorption and fluorescence spectra of biodiesel allowed to identify the molecules formed in the initial thermal degradation stage. Absorbance and fluorescence intensities systematically changed at thermal treatment over 100 °C. Therefore, monitoring of the degradation compounds allows to evaluate the biodiesel degradation and the evolution of the degradation compounds content. The results indicate that analytical methods based on UV–Vis absorption and fluorescence spectroscopy may be able to monitor the biodiesel degradation, providing basis for the development of simple, portable, and low-cost devices.

**Keywords** Biodiesel · UV–Vis absorption · Fluorescence · Thermal degradation

---

K. F. Magalhães (✉) · A. R. L. Caires · S. L. Oliveira
Grupo de Óptica e Fotônica, Instituto de Física, Universidade Federal de Mato Grosso do Sul, CP 549, Campo Grande, MS 79070–900, Brazil
e-mail: keurison@yahoo.com.br

T. A. Chimenez
Instituto Federal Catarinense, São Bento do Sul, SC CP 89283-064, Brazil

M. C. Fripp · F. Machado
Grupo de Óptica Aplicada, Faculdade de Ciências Exatas e Tecnologia, Universidade Federal da Grande Dourados, CP 533, Dourados, MS 79804–970, Brazil



M. Trindade (ed.), *Increased Biodiesel Efficiency*, Green Energy and Technology,
https://doi.org/10.1007/978-3-319-73552-8_5

# 5.1 Introduction

## *5.1.1 Biodiesel*

Biodiesel is an alternative fuel, environmentally friendly, and economically competitive compared to the fossil diesel (Van Gerpen 2005). Biodiesel is chemically defined as alkyl esters derived from vegetable oils or animal fats and is usually obtained by transesterification of triglycerides in the presence of an alcohol and an acid or basic catalyst, and glycerol as a by-product (Fig. 5.1) (Knothe et al. 2010).

Comparing the emissions of biodiesel and diesel, the combustion of biodiesel produces less soot than diesel because biodiesel is composed of esters and a small amount of aromatic compounds, resulting in an incomplete burning of fuel (Dantas et al. 2011). Physicochemical properties of biodiesel are influenced by the types and levels of fatty acids present in their composition (Fig. 5.2). It is well established that biodiesel is more susceptible to oxidation process than fossil diesel during transportation and storage due to the high level of unsaturation of the alkyl esters, unless its composition be altered by additives, which delay the oxidative process (McCormick et al. 2007; Navarra et al. 2011).

## *5.1.2 Biodiesel Stability*

Biodiesel quality during storage, transport, and use constitutes an important technical aspect to be evaluated since products of degradation may cause corrosion in the engine and clogging of filters and injection system (Singer and Rühe 2014). Hence, oxidation (formation of free radicals, arrangement with oxygen, formation and cleavage of peroxides in unsaturation, release of aldehydes, carboxylic acids, and polymer formation) is a relevant feature to be monitored in the biodiesel cycle because it contains esters of unsaturated fatty acids such as oleic, linoleic, and

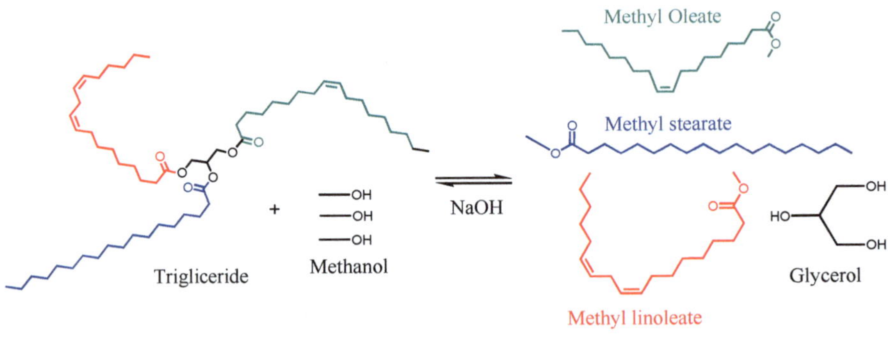


**Fig. 5.1** Transesterification scheme to biodiesel production

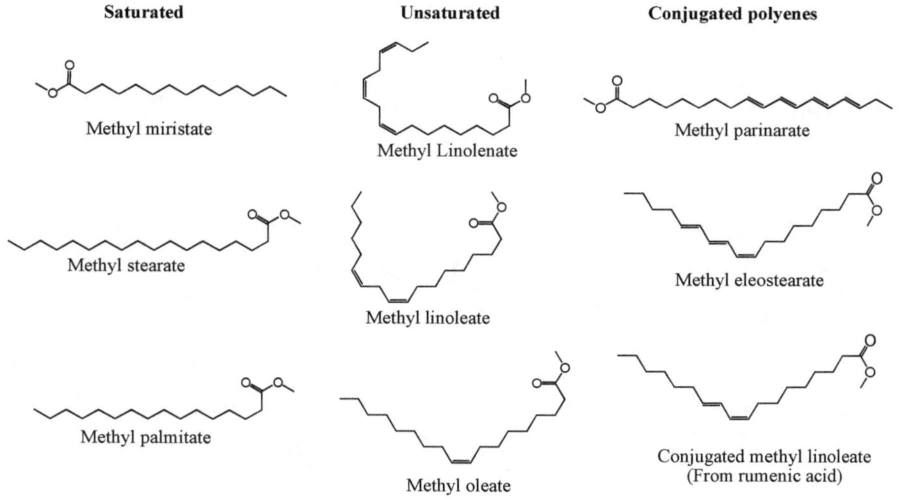


**Fig. 5.2** Methyl esters structure present in biodiesel

linolenic acids (Chacón et al. 2000; Dos Santos et al. 2011). These compounds under conditions of heating, ultraviolet radiation, humidity, atmospheric air, and presence of metals, even for a short time, may undergo oxidation by means of different mechanisms, for example, thermal oxidation, photo-oxidation, autoxidation, and hydrolytic reactions (Herbinet et al. 2008; Khoury et al. 2011; Atadashi et al. 2012; Jain and Sharma 2012; Magalhães et al. 2014; Yaakob et al. 2014).

In turn, biodiesel degradation can be delayed by using antioxidant agents, so increasing storage time and avoiding formation of by-products harmful to biodiesel and to engine. Synthetic antioxidants such as butyl-hydroxy-anisole (BHA), butyl-hydroxy-toluene (BHT), tert-Butylhydroquinone (TBHQ), and propyl gallate (PG) are the most widespread antioxidants in industry and they are known to retard the oxidation effects (increase in viscosity, acidity, and peroxide number). In addition to the synthetic antioxidants, it worth to point out that natural antioxidants such as tocopherol, chlorophyll, and carotenoids may also be present in biodiesel (Fig. 5.3) (Dunn 2005; Knothe 2007).

### *5.1.3 Standard Method to Monitor Biodiesel Stability*

Rancimat method is the standard method in Brazil for assessing oxidative stability of biodiesel at preset temperature in accordance with the European standard EN 14112 (Lapuerta et al. 2012). This method is also used to verify the effect of different heating on the thermal stability of biodiesel (Dunn 2008; Sarin et al. 2009; Xin et al. 2009; Karavalakis et al. 2010). In this approach, the oxidation stability is assessed by passing an air flow (oxygen) through the sample at a predetermined

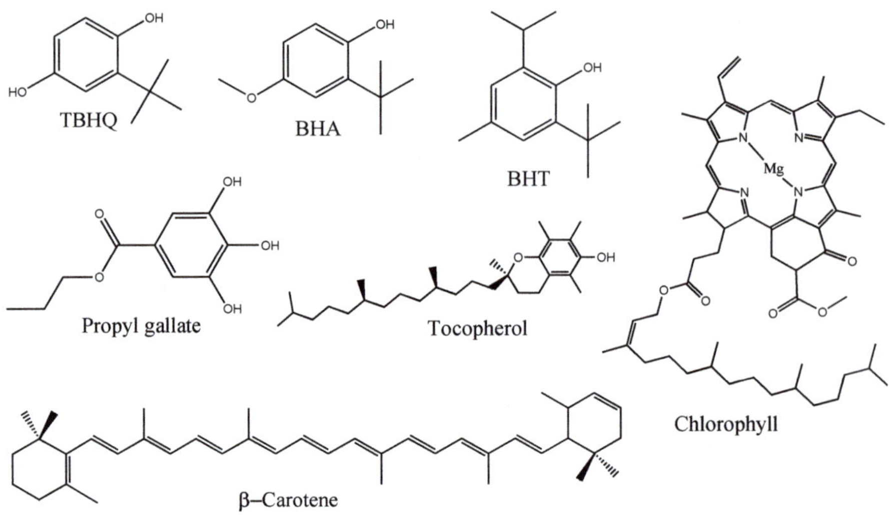


**Fig. 5.3** Synthetic and natural antioxidants present in biodiesel and vegetable oil

temperature; then, the induction time to change the maximum oxidation rate of the sample is determined, which is related to an increase in the conductivity of ultrapure water contained in a vessel (Fig. 5.4a). The heated biodiesel sample in the presence of oxygen starts to oxidize, forming organic acids of low molecular weight. Then, these acids leave the sample and are directed to the measuring cell, which allows to evaluate changes in the conductivity of the water promoted by these acids. Induction period is determined by the inflection point of the water conductivity curve as a function of time (Fig. 5.4b) (Knothe et al. 2010).

Despite the advantages presented by the Rancimat method, as simple operation and analysis, electrical conductivity profiles distinct from the one shown in

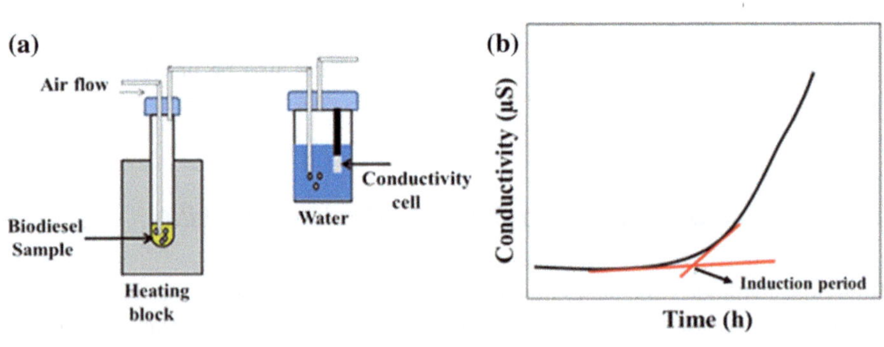


**Fig. 5.4** Rancimat experimental scheme (**a**) and typical curve of conductivity as a function of time to determine the induction period (**b**)

Fig. 5.4b could be achieved, hampering the proper determination of the induction period (Knothe et al. 2010).

### *5.1.4 Alternative Methods to Assess Biodiesel Stability*

Chromatographic methods can also be applied as alternative approach to analyze the degradation compounds from biodiesel (Leung et al. 2006; Jin et al. 2010). Nevertheless, these techniques are costly and require preseparation steps of the constituents prior to analysis. Thermal methods such as thermogravimetry (TG) and differential scanning calorimetry (DSC) have been also used in studies of the thermo-oxidative stability of biodiesel (Kivevele et al. 2011; Dunn 2012). Among the analytical approaches already reported, infrared absorption spectroscopy, UV–Vis absorption, and fluorescence were also indicated as potential methods for biodiesel stability monitoring (Conceição et al. 2007; Ferrari and De Souza 2009; de Lira et al. 2010; Meira et al. 2011; Magalhães et al. 2014). Although various techniques have been tested, studies have to be developed to identify alternative approaches capable of monitoring the degradation processes of biodiesel when submitted to thermal stress in a rapid and accurate way without the need for sample pretreatment. In this chapter, an alternative analytical tool for monitoring biodiesel thermal degradation in the early stages based on endogenous chromophores and fluorophores is discussed.

## 5.2 Evaluation of Absorption and Fluorescence Methods to Monitor Biodiesel Thermal Degradation

### *5.2.1 Biodiesel Production*

Soybean and canola biodiesel were obtained via alkaline transesterification of the respective refined oils of same brand using NaOH as alkaline catalyst (0.4 wt% with respect to oil) and 6:1 molar ratio of methanol/oil (Chimenez et al. 2012). Duplicate aliquots (200 g) of biodiesel were placed into beaker and heated for 12 h from room temperature to 210 °C. An oven without airflow was used to induce the early stages of biodiesel oxidation. During the heating process, the samples were maintained during 1 h at 40, 50, 70, 90, 100, 120, 135, 150, 170, 190, and 210 °C in which 5-mL portion was collected of each batch prior to the temperature change. Therefore, the samples underwent cumulative effects. Aliquots were stored in dark bottles until the UV–Vis absorption and fluorescence measurements. Unheated biodiesel samples were also collected for analysis.

### 5.2.2 Fluorescence and Absorption Measurements

Measurements of molecular absorption in the 200–400 nm range were carried out in a Cary 50 spectrophotometer (Varian). These measurements were made at room temperature using a quartz cuvette of 10-mm and 0.5-mm path length for the diluted and undiluted samples, respectively. Soybean and canola biodiesel samples diluted in n-hexane (HPLC grade) (around 1.1 mg/mL) were used to ensure the linear regime of the Beer–Lambert law and avoid inner filter effects in the fluorescence measurements (Lakowicz 2006; Divya and Mishra 2008; Magalhães et al. 2014).

Emission and excitation spectra (emission–excitation contour maps) of the samples were recorded at room temperature using a bench fluorescence spectrophotometer Cary Eclipse (Varian). Excitation wavelength was scanned between 250 and 400 nm, while the emission was monitored from 280 to 600 nm at intervals of 2 nm with excitation and emission slits of 10 nm for the diluted samples. Excitation/Emission contour maps from the undiluted samples were monitored from 370 to 700 nm at intervals of 2 nm with excitation and emission slits of 5 nm, while the excitation wavelength spanned from 300 to 420 nm. Fluorescence was measured at a 90° angle relative to the excitation light (right-angle geometry). All emission/excitation contour maps were recorded at room temperature using 10-mm path length quartz cells. The n-hexane used in dilution did not exhibit any fluorescence in the spectral range evaluated (data not shown).

Figure 5.5 shows the UV–Vis absorption spectra of the diluted canola and soybean biodiesel for all temperatures used in the thermal treatment. Absorption bands with maxima around 232, 260, 270, 280, 300, and 315 nm were observed for both samples, indicating that the same chromophores were responsible for the obtained absorptions in both canola and soybean biodiesel. However, the concentration of these chromophores in canola and soybean biodiesel is not the same since the absorption intensities for the corresponding spectral region are different. The absorption band around 232 nm could be attributed to compounds with two conjugated double bonds C=C (conjugated dienes, i.e., conjugated linoleic acid (Sklar et al. 1977a, b; Arudi et al. 1983). Absorption in the 260–290 nm range may be assigned to conjugated trienes (i.e., eleostearic acid) (Rekdal and Mel 1995; Fygle and Melo 1996; Smyk et al. 2009, 2011). In turn, the absorption bands at 300 and 315 nm are associated with conjugated tetraene compounds (i.e., parinaric acid) (Smyk et al. 2009, 2011; Magalhães et al. 2014).

The absorbance at around 232, 270, 300, and 315 nm increased in the canola and soybean biodiesel subjected to temperatures exceeding about 100 °C, as highlighted in Figs. 5.6 and 5.7. This increase in absorption could be related to the early stage of the thermal oxidation of methyl esters that results in changes in the position and number of carbon–carbon double bonds. As consequence, an escalation in the concentration of hydroperoxide with two conjugated double bonds C=C and products as conjugated trienes and tetraenes should be expected. Observed fluctuations in the absorbance values may be associated with the competition between

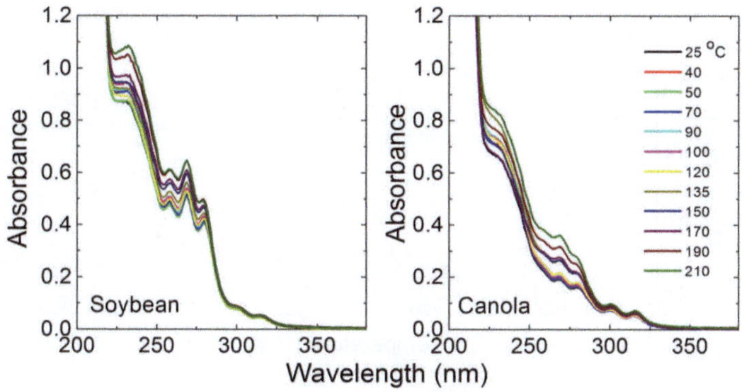


**Fig. 5.5** Absorbance spectra of the diluted soybean (left) and canola (right) biodiesel as a result of the temperature

the formation of absorbing species and the kinetics of lipid oxidation. Indeed, the monitoring of these chromophores in diluted biodiesel by UV–Vis absorption allows following the early evolution of the concentration of the degradation products resulting from the heating (Sklar et al. 1977a, b; Arudi et al. 1983; Rekdal and Mel 1995; Fygle and Melo 1996; Smyk et al. 2009, 2011; Magalhães et al. 2014).

Conjugated dienes, trienes, and tetraenes were also detected in the untreated biodiesels; hence, the vegetable oils used as raw materials already had in their compositions triglycerides with conjugated double bonds. The presence of these molecules could be explained by either the characteristics of the triglycerides in the oils or the triglyceride degradation during oil refining, storage, and distribution.

Fluorescence spectroscopy is a valuable tool in studies related to physical–chemical features of biodiesel (Scherer et al. 2011; Caires et al. 2012, 2014). Magalhães and coworkers have attributed the fluorescence from biodiesel in the 350–500 nm range the presence of conjugated tetraenes in its composition

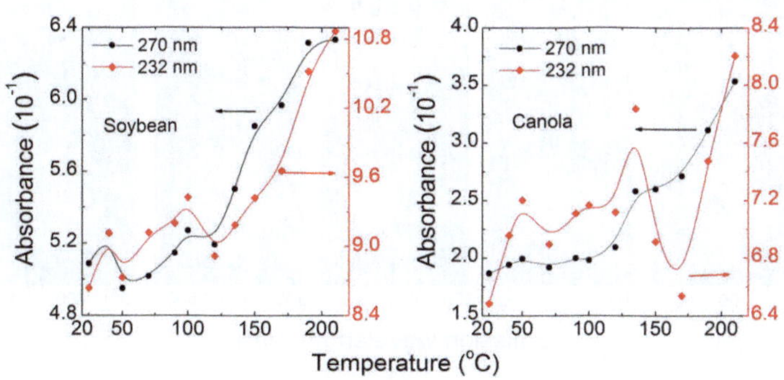


**Fig. 5.6** Absorbance at 232 and 270 nm for the diluted soybean (left) and canola (right) biodiesel as a function of temperature

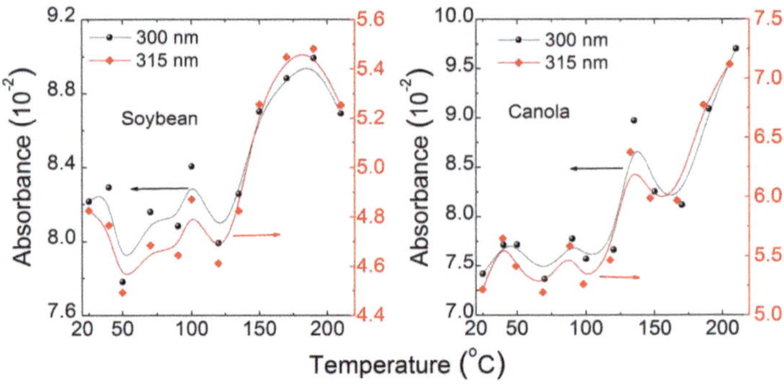


**Fig. 5.7** Absorbance at 300 and 315 nm for the diluted soybean (left) and canola (right) biodiesel as a function of temperature

(Magalhães et al. 2014). As UV–Vis absorption spectra reveal that the concentration of conjugated tetraenes increased as a consequence of the thermal degradation of the constituents of the biodiesel, excitation–emission contour maps were collected for all samples in order to evaluate whether these molecules could be used as a fluorescence probe to assess biodiesel degradation in an accelerated degradation test using an oven without air flow. Figure 5.8 displays representative excitation–emission contour maps of canola and soybean biodiesel submitted to different heating temperatures. Measurements were made using the same diluted samples

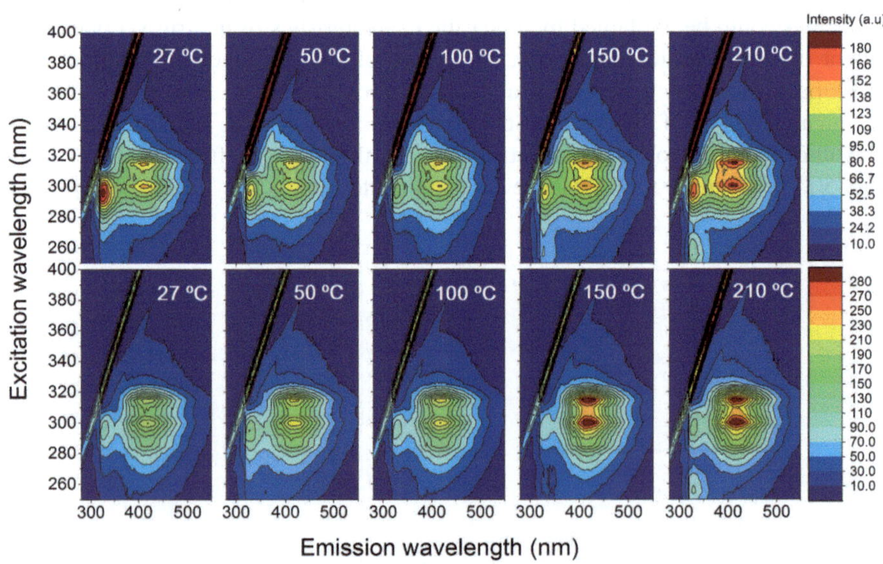


**Fig. 5.8** Excitation–emission contour maps for the diluted soybean (on top) and canola (at bottom) biodiesel in different temperatures. Rayleigh and Raman scatterings are not omitted

employed in the UV–Vis absorption experiments to allow comparative analysis. Raman and Rayleigh scatterings were not excluded from the spectra.

Contour maps of both soybean and canola biodiesel exhibit similar spectral profiles as a function of the heat treatment. Emission between 300 and 350 nm, characteristic of phenolic compounds, tocopherols and tocotrienols, was observed under excitation in the 270–310 nm range (Hurtubise 1976; Sikorska et al. 2004, 2005a, b; Dupuy et al. 2005; Zandomeneghi et al. 2005; Escuderos et al. 2009; Poulli et al. 2009; Ramalho et al. 2011). Two other emission regions between 350 and 550 nm were detected with excitation around 300 and 315 nm, which could be assigned to conjugated tetraenes (Smyk et al. 2009; Magalhães et al. 2014). Figure 5.8 shows also differences in the fluorescence intensities of biodiesel samples subjected to heating at temperatures above 100 °C. To evaluate these spectral changes in the fluorescence of diluted soybean and canola biodiesel, Fig. 5.9 presents the relationship of fluorescence intensity at 420 nm of samples under 300 nm excitation and heating temperature.

Fluorescence intensity of diluted soybean and canola biodiesel presented similar dependence with the treatment temperature when excited at 300 nm. A remarkable increase in the fluorescence intensity at 420 nm with the increasing temperature was observed for temperatures over 100 °C. Similar tendency was also observed for fluorescence intensity at 420 nm of samples submitted to 315 nm excitation (data not shown). As previously stated, this excitation–emission spectral range has been attributed to conjugated tetraenes (Smyk et al. 2009; Magalhães et al. 2014).

Consequently, these fluorescence data corroborate the findings based on UV–Vis absorption measurements which indicated an increase in the concentration of conjugated tetraenes in biodiesel as a result of the heat treatment. Therefore, evaluation of conjugated tetraenes in diluted biodiesel by UV–Vis absorption and/or fluorescence measurements may be useful for monitoring the initial degradation processes of compounds present in biodiesel.

In addition to the analysis performed with diluted samples of biodiesel, UV–Vis absorption and fluorescence measurements were also carried out with undiluted

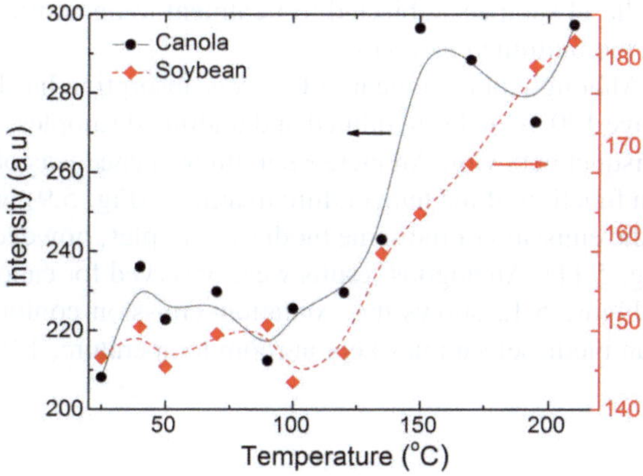


**Fig. 5.9** Fluorescence intensity at 420 nm for the diluted soybean and canola biodiesel under 300 nm excitation as a function of temperature

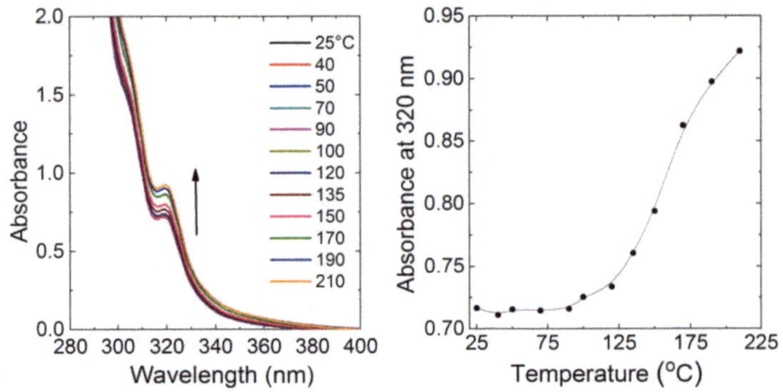


**Fig. 5.10** Absorbance spectra of undiluted soybean biodiesel and absorbance at 320 nm as a function of the temperature treatment

biodiesel samples (as-made) to evaluate whether dilution would be a fundamental step of sample preparation to monitor degradation of biodiesel constituents by means of such techniques. Figure 5.10 shows UV–Vis absorption spectra of the undiluted samples submitted to different treatment temperatures of the thermal treatment. Absorptions shoulders at around 300 and 320 nm were detected in the absorption spectra, which can be attributed to conjugated tetraenes, as previously discussed (Smyk et al. 2009; Magalhães et al. 2014). A significant increase in absorption of these samples subjected to heating at temperatures above 100 °C is also evident. This behavior may be associated with the formation of new compounds due to the degradation of the biodiesel constituents, producing conjugated tetraenes. Similar results were obtained for undiluted canola biodiesel (data not shown).

Figure 5.11 displays fluorescence spectra of the undiluted soybean biodiesel samples, under excitation at 350 nm, with regards to the different treatment temperatures. The spectra show emission in the spectral range between 350 and 550 nm; this fluorescence band has been attributed to tetraene conjugates in biodiesel (Magalhães et al. 2014). In addition, a decrease in the fluorescence of the biodiesel samples subjected to treatment temperatures above 100 °C was observed for the undiluted samples.

Although an increment in UV–Vis absorption has been observed at temperatures above 100 °C in both, diluted and undiluted samples, fluorescence intensity showed a distinct behavior. An increase in fluorescence was observed in the diluted samples as a function of the temperature treatment (Fig. 5.9), while a reduction was detected in the emission of the same biodiesel samples, however undiluted for measurements (Fig. 5.11). Analogous results were observed for canola biodiesel (data not shown).

Figure 5.12 shows the excitation–emission contour maps of the undiluted soybean biodiesel samples kept at room temperature, 135 and 210 °C. Each one of the

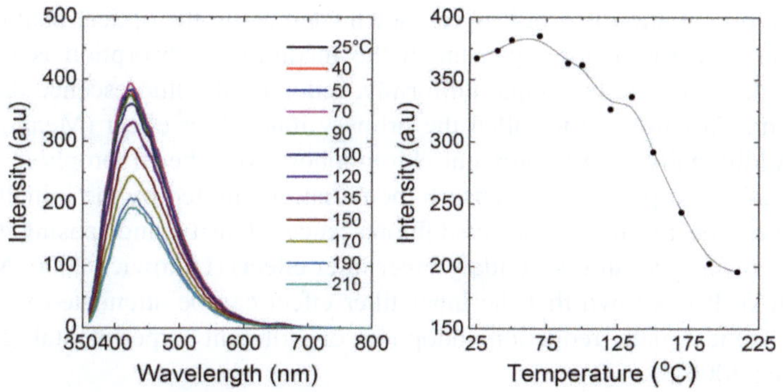


**Fig. 5.11** Fluorescence spectra of undiluted soybean biodiesel and fluorescence intensity at 430 nm as a function of the temperature treatment

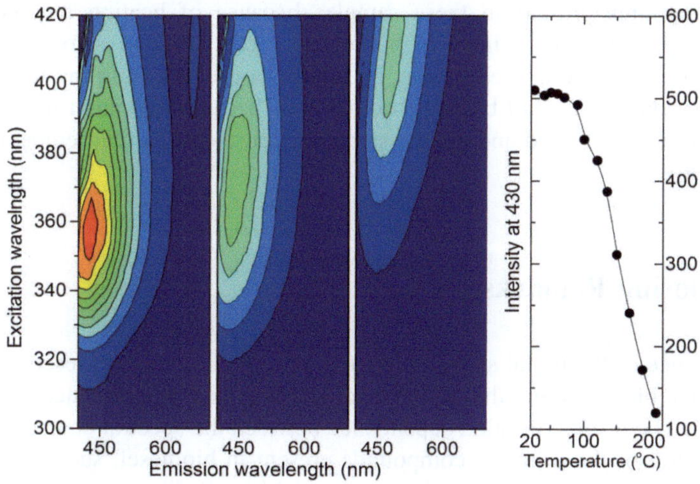


**Fig. 5.12** Excitation/emission contour maps of the undiluted soybean biodiesel for different temperatures and fluorescence intensity at 430 nm of soybean biodiesel under 360 nm excitation as a function of temperature treatment

maps reveals only one excitation–emission region, differently from the observed in diluted biodiesel which exhibited three regions (Fig. 5.8). A systematic decrease in the fluorescence intensity as well as a redshift of the contour maps can be also observed as a result of the increase in the temperature treatment, as discussed previously.

Radiation reabsorption process can occur during fluorescence measurements in undiluted samples, causing distortions in the absorption and fluorescence spectra. This process has been called inner filter effect and may involve both excitation and emission radiation (Lakowicz 2006; Mendonça et al. 2013).

Excitation radiation is progressively attenuated along the optical path due to the absorption of chromophores present in the medium. If absorption is significant, radiation does not excite sample uniformly, reducing the fluorescence detected by the system. This process is called the primary inner filter effect (Mendonça et al. 2013). Additionally, non-fluorescent chromophores or other fluorophores of either same or different species can reabsorb the radiation emitted by the initially excited fluorophores, decreasing the detected fluorescence intensity, and causing distortions in the emission spectrum (secondary inner filter effect) (Lakowicz 2006; Mendonça et al. 2013). It is known that the inner filter effect can be attenuated with sample dilution, optical path reduction, adoption of different experimental geometries (Lakowicz 2006).

The results shown in Fig. 5.12 are quite distinct of those obtained in diluted samples, revealing the strong impact of the inner filter effect on the spectral profiles (Lakowicz 2006; Mendonça et al. 2013). This corroborates the need of dilution of the sample to properly evaluate the evolution of the concentration of chromophores and fluorophores present in these samples because of heating treatment. Even though inner filter effect impacts fluorescence of undiluted samples under the experimental conditions used, it may be used as a useful tool in the monitoring of initial degradation steps of biodiesel compounds because alteration in fluorescence intensity of the undiluted and diluted samples occurred in similar temperatures (Gil et al. 2006).

## 5.3 Closing Remarks

The influence of the initial stages of degradation of compounds present in soybean and canola biodiesel on the UV–Vis absorption and fluorescence spectra was evaluated. The results revealed dependence between the spectral characteristics and concentration of the different compounds present in biodiesel, such as dienes, trienes, and conjugated tetraenes. It was also evidenced that at temperatures above 100 °C, soybean and canola biodiesel showed changes in the UV–Vis absorption and fluorescence spectra for both diluted and undiluted samples. Thus, it is possible to monitor the initial stages of biodiesel degradation via both absorption and fluorescence measurements. In summary, the findings indicate that UV–Vis absorption and fluorescence may be used in the development of new methodologies and devices for assessing physical–chemical alteration in biodiesel composition due to the temperature effect.

**Acknowledgements** The authors are grateful for financial support from Conselho Nacional de Desenvolvimento Científico e Tecnológico (CNPq), Coordenação de Aperfeiçoamento de Pessoal de Nível Superior (CAPES), and Fundação de Apoio ao Desenvolvimento do Ensino, Ciência e Tecnologia do Estado de Mato Grosso do Sul (FUNDECT). This work was performed under the auspices of the National Institute of Science and Technology of Photonics/CNPq. One of the authors (A.R.L.C.) also appreciates the support provided by National Institute of Science and Technology of Optics and Photonics/CNPq.

## References

Arudi RL, Sutherland MW, Bielski BHJ (1983) Purification of oleic acid and linoleic acid. J Lipid Res 24:485–488

Atadashi IM, Aroua MK, Abdul Aziz AR, Sulaiman NMN (2012) The effects of water on biodiesel production and refining technologies: a review. Renew Sustain Energy Rev 16:3456–3470

Caires ARL, Lima VS, Oliveira SL (2012) Quantification of biodiesel content in diesel/biodiesel blends by fluorescence spectroscopy: evaluation of the dependence on biodiesel feedstock. Renew Energy 46:137–140. https://doi.org/10.1016/j.renene.2012.03.026

Caires ARL, Scherer MD, De Souza JE et al (2014) The role of viscosity in the fluorescence behavior of the diesel/biodiesel blends. Renew Energy 63:388–391. https://doi.org/10.1016/j.renene.2013.09.041

Chacón JN, Gaggini P, Sinclair RS, Smith FJ (2000) Photo- and thermal-oxidation studies on methyl and phenyl linoleate: anti-oxidant behaviour and rates of reaction. Chem Phys Lipids 107:107–120. https://doi.org/10.1016/S0009-3084(00)00157-2

Chimenez TA, Magalhães KF, Caires ARL, Oliveira SL (2012) Fluorescence as an analytical tool for assessing the conversion of oil into biodiesel. J Fluoresc 22:1177–1182. https://doi.org/10.1007/s10895-012-1057-x

Conceição MM, Fernandes VJ, Araújo AS et al (2007) Thermal and oxidative degradation of castor oil biodiesel. Energy Fuels 21:1522–1527. https://doi.org/10.1021/ef0602224

Dantas MB, Albuquerque AR, Barros AK et al (2011) Evaluation of the oxidative stability of corn biodiesel. Fuel 90:773–778. https://doi.org/10.1016/j.fuel.2010.09.014

de Lira LFB, de Albuquerque MS, Pacheco JGA et al (2010) Infrared spectroscopy and multivariate calibration to monitor stability quality parameters of biodiesel. Microchem J 96:126–131. https://doi.org/10.1016/j.microc.2010.02.014

Divya O, Mishra AK (2008) Understanding the concept of concentration-dependent red-shift in synchronous fluorescence spectra: prediction of lambda (SFS)(max) and optimization of deltalambda for synchronous fluorescence scan. Anal Chim Acta 630:47–56. https://doi.org/10.1016/j.aca.2008.09.056

Dos Santos VML, Da Silva JAB, Stragevitch L, Longo RL (2011) Thermochemistry of biodiesel oxidation reactions: a DFT study. Fuel 90:811–817. https://doi.org/10.1016/j.fuel.2010.09.017

Dunn RO (2005) Effect of antioxidants on the oxidative stability of methyl soyate (biodiesel). Fuel Process Technol 86:1071–1085. https://doi.org/10.1016/j.fuproc.2004.11.003

Dunn RO (2008) Effect of temperature on the oil stability index (OSI) of biodiesel. Energy Fuels 22:657–662. https://doi.org/10.1021/ef700412c

Dunn RO (2012) Thermal oxidation of biodiesel by pressurized differential scanning calorimetry: effects of heating ramp rate. Energy Fuels 26:6015–6024. https://doi.org/10.1021/ef3010543

Dupuy N, Le Dréau Y, Ollivier D et al (2005) Origin of French virgin olive oil registered designation of origins predicted by chemometric analysis of synchronous excitation-emission fluorescence spectra. J Agric Food Chem 53:9361–9368. https://doi.org/10.1021/jf051716m

Escuderos ME, Sayago A, Morales MT, Aparicio R (2009) Evaluation of alpha-tocopherol in virgin olive oil by a luminiscent method. Grasas Aceites 60:336–342. https://doi.org/10.3989/gya.108308

Ferrari RA, De Souza WL (2009) Avaliação da estabilidade oxidativa de biodiesel de óleo de girassol com antioxidantes. Quim Nova 32:106–111. https://doi.org/10.1590/S0100-40422009000100020

Fygle KE, Melo TB (1996) Optical absorption studies of the kinetics of UV- and self-initiated autoxidation of linoleate micelles. Chem Phys Lipids 79:39–46. https://doi.org/10.1016/0009-3084(95)02505-7

Gil DB, De La Peña AM, Arancibia JA et al (2006) Second-order advantage achieved by unfolded-partial least-squares/residual bilinearization modeling of excitation-emission fluorescence data presenting inner filter effects. Anal Chem 78:8051–8058. https://doi.org/10.1021/ac061369v

Herbinet O, Pitz WJ, Westbrook CK (2008) Detailed chemical kinetic oxidation mechanism for a biodiesel surrogate. Combust Flame 154:507–528. https://doi.org/10.1016/j.combustflame.2008.03.003

Hurtubise RJ (1976) Selective fluorescence quenching and determination of phenolic antioxidants. Anal Chem 48:2092–2094

Jain S, Sharma MP (2012) Correlation development between the oxidation and thermal stability of biodiesel. Fuel 102:354–358. https://doi.org/10.1016/j.fuel.2012.06.110

Jin F, Zhong H, Cao J et al (2010) Oxidation of unsaturated carboxylic acids under hydrothermal conditions. Bioresour Technol 101:7624–7634. https://doi.org/10.1016/j.biortech.2010.04.056

Karavalakis G, Stournas S, Karonis D (2010) Evaluation of the oxidation stability of diesel/biodiesel blends. Fuel 89:2483–2489. https://doi.org/10.1016/j.fuel.2010.03.041

Khoury RR, Ebrahimi D, Hejazi L et al (2011) Degradation of fatty acid methyl esters in biodiesels exposed to sunlight and seawater. Fuel 90:2677–2683. https://doi.org/10.1016/j.fuel.2011.03.041

Kivevele TT, Mbarawa MM, Bereczky Á, Zöldy M (2011) Evaluation of the oxidation stability of biodiesel produced from moringa oleifera oil. Energy Fuels 25:5416–5421. https://doi.org/10.1021/ef200855b

Knothe G (2007) Some aspects of biodiesel oxidative stability. Fuel Process Technol 88:669–677

Knothe G, Krahl J, Van Gerpen J (2010) The biodiesel handbook, 2nd edn. Elsevier, Amsterdam

Lakowicz JR (2006) Principles of fluorescence spectroscopy

Lapuerta M, Rodríguez-Fernández J, Ramos A, Álvarez B (2012) Effect of the test temperature and anti-oxidant addition on the oxidation stability of commercial biodiesel fuels. Fuel 93:391–396. https://doi.org/10.1016/j.fuel.2011.09.011

Leung DYC, Koo BCP, Guo Y (2006) Degradation of biodiesel under different storage conditions. Bioresour Technol 97:250–256. https://doi.org/10.1016/j.biortech.2005.02.006

Magalhães KF, Caires ARL, Silva MS et al (2014) Endogenous fluorescence of biodiesel and products thereof: investigation of the molecules responsible for this effect. Fuel 119:120–128. https://doi.org/10.1016/j.fuel.2013.11.024

McCormick RL, Ratcliff M, Moens L, Lawrence R (2007) Several factors affecting the stability of biodiesel in standard accelerated tests. Fuel Process Technol 88:651–657. https://doi.org/10.1016/j.fuproc.2007.01.006

Meira M, Quintella CM, Tanajura ADS et al (2011) Determination of the oxidation stability of biodiesel and oils by spectrofluorimetry and multivariate calibration. Talanta 85:430–434. https://doi.org/10.1016/j.talanta.2011.04.002

Mendonça A, Rocha AC, Duarte AC, Santos EBH (2013) The inner filter effects and their correction in fluorescence spectra of salt marsh humic matter. Anal Chim Acta 788:99–107. https://doi.org/10.1016/j.aca.2013.05.051

Navarra G, Cannas M, D'Amico M et al (2011) Thermal oxidative process in extra-virgin olive oils studied by FTIR, rheology and time-resolved luminescence. Food Chem 126:1226–1231. https://doi.org/10.1016/j.foodchem.2010.12.010

Poulli KI, Chantzos NV, Mousdis GA, Georgiou CA (2009) Synchronous fluorescence spectroscopy: tool for monitoring thermally stressed edible oils. J Agric Food Chem 57:8194–8201. https://doi.org/10.1021/jf902758d

Ramalho EFSM, Santos IMG, Maia AS et al (2011) Thermal characterization of the poultry fat biodiesel. J Therm Anal Calorim 106:825–829. https://doi.org/10.1007/s10973-011-1886-x

Rekdal K, Mel TB (1995) UV-initiated autoxidation of methyl linoleate in micelles studied by optical absorption. Chem Phys Lipids 75:127–136. https://doi.org/10.1016/0009-3084(94)02413-Y

Sarin A, Arora R, Singh NP et al (2009) Influence of metal contaminants on oxidation stability of Jatropha biodiesel. Energy 34:1271–1275. https://doi.org/10.1016/j.energy.2009.05.018

Scherer MD, Oliveira SL, Lima SM et al (2011) Determination of the biodiesel content in diesel/biodiesel blends: a method based on fluorescence spectroscopy. J Fluoresc 21:1027–1031

Sikorska E, Gliszczyńska-Świgło A, Khmelinskii I, Sikorski M (2005a) Synchronous fluorescence spectroscopy of edible vegetable oils. Quantification of tocopherols. J Agric Food Chem 53:6988–6994. https://doi.org/10.1021/jf0507285

Sikorska E, Górecki T, Khmelinskii IV et al (2005b) Classification of edible oils using synchronous scanning fluorescence spectroscopy. Food Chem 89:217–225. https://doi.org/10.1016/j.foodchem.2004.02.028

Sikorska E, Romaniuk A, Khmelinskii IV et al (2004) Characterization of edible oils using total luminescence spectroscopy. J Fluoresc 14:25–35

Singer P, Rühe J (2014) On the mechanism of deposit formation during thermal oxidation of mineral diesel and diesel/biodiesel blends under accelerated conditions. Fuel 133:245–252. https://doi.org/10.1016/j.fuel.2014.04.041

Sirkoska E, Romaniuk A, Khmelinskii IV et al (2004) Characterization of edible oils using total luminescence spectroscopy. J Fluoresc 14:25–35. https://doi.org/10.1023/B:JOFL.0000014656.75245.62

Sklar LA, Hudson BS, Petersen M, Diamond J (1977a) Conjugated polyene fatty acids on fluorescent probes: spectroscopic characterization. Biochemistry 16:813–819. https://doi.org/10.1021/bi00624a001

Sklar LA, Hudson BS, Simoni RD (1977b) Conjugated polyene fatty acids as fluorescent probes: binding to bovine serum albumin. Biochemistry 16:5100–5108. https://doi.org/10.1021/bi00642a024

Smyk B, Amarowicz R, Szabelski M et al (2009) Steady-state and time-resolved fluorescence studies of stripped Borage oil. Anal Chim Acta 646:85–89. https://doi.org/10.1016/j.aca.2009.05.007

Smyk B, Wieczorek P, Zadernowski R (2011) A method of concentration estimation of trienes, tetraenes, and pentaenes in evening primrose oil. Eur J Lipid Sci Technol 113:592–596. https://doi.org/10.1002/ejlt.201000418

Van Gerpen J (2005) Biodiesel processing and production. Fuel Process Technol 86:1097–1107. https://doi.org/10.1016/j.fuproc.2004.11.005

Xin J, Imahara H, Saka S (2009) Kinetics on the oxidation of biodiesel stabilized with antioxidant. Fuel 88:282–286. https://doi.org/10.1016/j.fuel.2008.08.018

Yaakob Z, Narayanan BN, Padikkaparambil S et al (2014) A review on the oxidation stability of biodiesel. Renew Sustain Energy Rev 35:136–153. https://doi.org/10.1016/j.rser.2014.03.055

Zandomeneghi M, Carbonaro L, Caffarata C (2005) Fluorescence of vegetable oils: olive oils. J Agric Food Chem 53:759–766. https://doi.org/10.1021/jf048742p

# Chapter 6
# Recent Advances on Sample Preparation Procedures for Elemental Determination in Biodiesel

**Liriana M. Roveda, Marcela Z. Corazza and Jorge L. Raposo Jr.**

**Abstract** In recent years, renewable oilseeds have been investigated as viable alternatives sources for enhanced production of biofuels. However, due to the recent investigation for new energy sources, greater attention is being given to ensure quality of the final products, in particular to the elemental content, to increase the diversity of biodiesel. The elemental determination of biodiesel is often performed by spectrometric techniques, and most of them require a sample pre-treatment to match the sample characteristics (i.e., high organic matter content, physicochemical properties, and low concentrations of analyte) to the analytical techniques available. Moreover, due to the advances on environment-friendly procedures (based on the principles of green chemistry), alternative sample pre-treatments based on nondestructive measurement have been proposed, not only for the determination of metals in biodiesel, but also in the entire biodiesel production chain. By the way, it is important to emphasize that a nondestructive sample preparation procedure does not involve the complete decomposition of the matter by concentrated acid and heating. The use of nondestructive sample preparation in oily matrices for the determination of metals by atomic spectrometry is still a recent practice. Some of these are based on extraction procedures or by the formation of emulsion and microemulsion stable solutions. In many cases, these procedures can be performed using ultrasonic energies, thermostatic bath, centrifuge, vortex stirring, shaker, and other apparatus which improves the efficiency of the sample preparation procedure selected.

**Keywords** Biodiesel · Nondestructive methods · Sample preparation and metals

---

L. M. Roveda · M. Z. Corazza
Faculty of Exact Sciences and Technology, Federal University of Grande Dourados, 364 Highway Dourados-Itahum, Km 12, 79804-970 Dourados, MS, Brazil

J. L. Raposo Jr. (✉)
Institute of Chemistry, Federal University of Mato Grosso Do Sul, 549, 79070-900 Campo Grande, MS, Brazil
e-mail: jorgeluizraposo@gmail.com

 
M. Trindade (ed.), *Increased Biodiesel Efficiency*, Green Energy and Technology,
https://doi.org/10.1007/978-3-319-73552-8_6

## 6.1 Introduction

Nowadays, the presence of metals in fuels, unless they have been purposely added, is generally undesirable, since its excess may impair engine performance and cause oxidation of their parts (Saint'Pierre et al. 2004). In addition, the presence of metals is an environmental problem, since they can be released to atmosphere during the combustion of the fuel, damaging air quality. For biodiesel, it is known that metals exert a deleterious effect on the quality of biofuel, because even in trace amounts these elements act as catalysts in oxidation processes, reducing their oxidative stability and impairing their performance.

Metals can be incorporated into biodiesel during the production process, due to the contact of the oil with the equipment used in reactors and distillation units, and with metallic parts of the tanks used for storage. In addition, metals are naturally present in vegetable oils used as raw material for biodiesel production, due to their absorption from soil and water during the plant growth (Pereira et al. 2013). Metals such as Na and K, used as catalysts in the biodiesel production, may be present in the final product as abrasive solids and insoluble soaps, which can contribute to the wear of automotive engine parts (e.g., injector, fuel pump, pistons, rings). Hard water and drying agents such as $MgSO_4$ and CaO used in cleaning processes can contribute to increase the Ca and Mg levels in biodiesel. These elements can also form soaps, incorporate undesirable compounds into the engine, lead to injection and clogging faults of some engine parts (Lyra et al. 2010; Pereira et al. 2014). Storage materials composed of brass or bronze contribute to contaminate the biodiesel with Cu, Zn, Sn, and Pb, due to its interaction with biodiesel, resulting in insoluble sediments (Yaakob et al. 2014).

In recent years, many efforts have been directed in the search for effective and low-cost antioxidant products for biodiesel. However, monitoring the content of metals and other parameters that influence the oxidation of biofuel is still an essential step to ensure the quality of the final products. Governments and nationals agencies around the world have established standard norms that regulate the quality of biodiesel (B100) by describing requirements and control methods. For instance, the European Standard EN 14214, the American Standard ASTM D 6751, and other standards, such as the Brazilian Standard RANP 07/08 and Argentine Standard IRAM 6515-1:2006 establish the maximum of 5.0 mg $kg^{-1}$ for Na + K, as well as for Ca + Mg (Knothe 2006; Avila Orozco et al. 2014). By the way, the limits for transition metals are not established yet, although they act as strong catalysts in the oxidation of biodiesel, even those in trace concentration (Knothe and Dunn 2003; Sarin et al. 2009, 2010; Jain and Sharma 2011).

The elemental determination in any type of sample requires the availability of sensitive analytical techniques that allows simultaneous or at least sequential determinations. Although standard norms such as EN 14214 recommends the use of flame atomic absorption spectrometry (FAAS) for the determination of Na and K (analytical methods EN 14108 and EN 14109) and inductively coupled plasma atomic emission spectrometry (ICP-OES) for determination of Ca and Mg (method EN 14538) (Knothe 2006), several methods to determine these and other metals in biodiesel have been described in the literature during the last years using other analytical techniques.

Despite the variety of techniques that have been successfully used in elemental analyses in different areas of analytical chemistry, there is no doubt that the spectrometric techniques (e.g., ICP-OES, ICP-MS, AAS, AES, and AFS) still figure as most used for determination of metals in biodiesel (Edlund et al. 2002; dos Santos et al. 2007; Woods and Fryer 2007; Chaves et al. 2008a, b; de Jesus et al. 2008; de Souza et al. 2008; Aranda et al. 2009; Lobo et al. 2009, 2011; Iqbal et al. 2010; Amais et al. 2010; Lyra et al. 2010; Quadros et al. 2011; Barros et al. 2012; Pillay et al. 2012; Pereira et al. 2013, 2014; Raposo et al. 2014).

It should be noted that each analytical technique available for elemental analysis has advantages and limitations; in others words, there is no a perfect technique, and it is up to the analyst to choose the one that is best suited for a specific analysis. However, whatever the analytical technique adopted, the sample should initially be subjected to a pretreatment step which covers a series of operations that are necessary to modify chemical and/or physical properties of the sample to make it compatible with the analytical technique, or improve the performance of the analysis. There are few commercial instruments that do not require or require the least amount of sample preparation, but the lack of a reliable calibration method is a common problem. In many cases, the purpose of the sample preparation steps includes dissolution of the analyte in a suitable solvent or concentrated acid, the isolation of the target analyte from the interfering compounds, and preconcentration of the analyte (Bader and Zimmermann 2012).

## 6.2 Nondestructive Sample Preparation Procedures for Biodiesel

It is consensual that sample preparation is the most critical step of the chemical analysis, since it consumes about 60% of the analytical work and contributes to 30% of the errors of the analysis (Oliveira 2003). Therefore, this step must be carefully conducted because it depends on the precision and accuracy of the results as well as the total time and costs involved.

The choice of the adequate sample preparation procedure should consider the target analyte and its concentration, sample matrix, instrumental technique, amount of sample required, precision, and accuracy. In general, routine sample preparation procedures are well described in the literature. In other cases, it will be necessary to develop a new, reliable, and applicable sample preparation procedure. However, procedures with low reagent consumption (especially acids and toxic or aggressive solvents to the environment), low time consuming, and minimal sample handling are the most desirable.

Concerning the complex samples such as biodiesel, the challenge of making them suitable for the available analytical technique is large, mainly due to high organic matter, physical characteristics of the matrix, water immiscibility, and low concentrations of the target elements (Nunes et al. 2011). However, nondestructive sample preparation procedures for biodiesel analysis had a major breakthrough due to the advances in environment-friendly procedures (based on the principles of green chemistry). Although most of the papers report the use of the conventional decomposition by concentrated acid and heating on the sample preparation step, there is an increasing trend focus on the nondestructive sample preparation procedures field. Most of these procedures are based on direct dilution of the samples with organic solvents, emulsion and microemulsion solutions, extraction induced by emulsion breaking, and liquid–liquid extractions. Some of these procedures reported in the literature for elemental analysis in biodiesel are summarized in Table 6.1.

Based on this and considering the importance of monitoring the metal content in biodiesel, the goal of this chapter is to describe some potential nondestructive samples preparation for elemental analysis of biodiesel in recent years, especially combined with spectrometric techniques, and some other traditional sample preparations procedures routinely used.

## *6.2.1 Sample Dilution*

Traditional sample preparation procedures used for elemental determination in biodiesel are based on direct dilution with organic solvents (dos Santos et al. 2007; Quadros et al. 2011; Chaves et al. 2011; Barros et al. 2012). Xylene, isobutylmethylketone, dimethylbenzene, 1,2-dimethylbenzene, kerosene, pentane, hexane, and heptane are the most common solvents used (Hardaway et al. 2004). The European Standards (BS EN 14108/2003; BS EN 14109/2003; DIN EN 14538/2006), and also the American (ASTM D6751-10 2010), and Brazilian (ABNT NBR 15553:2008; ABNT NBR 15556:2008) norms recommend the dilution of biodiesel samples in xylene or kerosene for the determination of Na, K, Mg, and Ca by spectrometric techniques. In these cases, the calibrations are performed with organometallic standards in base oil dissolved in the same solvent as the samples.

Ghisi et al. (2011) reported the use of tetramethylammonium hydroxide in the dilution of biodiesel samples for the determination of Cu and Fe by electrothermal atomic absorption spectrometry. The sample was prepared by mixing 500 mg of the biodiesel with 500 μL of the 25% (v/v) TMAH solution. After mixing, the flask was heated to 90 °C in a hot plate for 5 min. The solution was diluted to 5.0 mL with water. According to the authors, the solution was not clear for some samples, but the final slurry was enough stable and homogeneous to allow its measurement. The calibration curves were obtained by dissolutions of the analytes in TMAH medium.

Despite the simplicity, many drawbacks are related to the application of this preparation procedure, such as toxicity of the solvents, high cost of organometallic

**Table 6.1** Application of nondestructive methods for elemental determination in biodiesel samples

| Biodiesel | Element | Analytical technique | Sample preparation | Solutions mixture | Calibration | DL | References |
|---|---|---|---|---|---|---|---|
| Cottonseed, fodder turnip, castor, soybean, soy sludge, and frying oil | Al, Cu, Fe, and Mn | ETAAS | Dilution | $HNO_3$, ethanol | External calibration with aqueous standards in ethanol (Al) and external calibration against aqueous solutions (Cu, Fe, and Mn) | NR | Quadros et al. (2011) |
| NR | Si | MIP-OES | Dilution | Ethanol | External calibration in aqueous solution | 20 μg $L^{-1}$ | Amais et al. (2013) |
| Soybean, sunflower, cotton and canola and animal sources | Na and K | FAES | Dilution | Ethanol; biodiesel | External calibration in biodiesel/ethanol solution | 0.65–1.20 mg $kg^{-1}$ | Barros et al. (2012) |
| Animal fat, castor bean, fodder turnip, rapeseed, soybean, and residual cook oil | Ca, Cu, Fe, K, Mg, Na, P, S, and Zn | ICP-OES | Dilution | $HNO_3$, ethanol or 1-propanol | External calibration with internal standardization (Y) in ethanol or 1-propanol | 0.001–0.4 μg $g^{-1}$ | Chaves et al. (2011) |
| NR | Na, K, Cr, and V | WCAES | Dilution | Methanol or ethanol | Standard additions (Na, K, and Cr) and external calibration (V) | 20–90 μg $kg^{-1}$ | Dancsak et al. (2014) |
| NR | Ca and Mg | FAAS | Dilution | Ethanol, $HNO_3$ | Internal standard addition (Sr and Mn) | NR | Fortunato et al. (2015) |
| Cottonseed, castor bean, fodder turnip, soybean, | Cu and Fe | ETAAS | Dilution | Tetramethylammonium hydroxide (TMAH), water | External calibration with standards treated as samples | 15 and 24 ng $g^{-1}$ | Ghisi et al. (2011) |

(continued)

**Table 6.1** (continued)

| Biodiesel | Element | Analytical technique | Sample preparation | Solutions mixture | Calibration | DL | References |
|---|---|---|---|---|---|---|---|
| soy sludge, frying oil and animal fat | | | | | | | |
| Soybean, canola, sunflower, cotton, macaw, residual oil, tallow, and biodiesel blends | Na, K and Ca | FAES | Emulsion | $HNO_3$, n-butanol, Triton X-100, water | External calibration whit internal standardization (Li) | 0.005–0.3 mg $kg^{-1}$ | Raposo et al. (2014) |
| Soya | Cd and Hg | ETAAS | Emulsion | $HNO_3$, Triton X-100, isopropyl alcohol, water | Analyte addition | 0.3 and 10.2 µg $kg^{-1}$ | Aranda et al. (2012) |
| Cotton, soybean lees, castor bean, waste frying, and soybean | Co, Cu, Fe, Mn, Ni, and V | ICP-MS | Emulsion | Triton X-100, $HNO_3$, water | External calibration whit internal standardization ($^{103}$Rh) | 0.3–1.5 ng $g^{-1}$ | Chaves et al. (2008a) |
| Soybean, algae, and castor | Ca, Cu, Fe, Mn, Mg, Na, K, and Si | ICP-OES | Emulsion | Formic acid, Triton X-100, antifoam, water | External calibration with inorganic aqueous standards solution in the emulsion medium | 0001–0241 mg $kg^{-1}$ | Lisboa et al. (2013) |
| Soya | Hg | AFS | Emulsion | $HNO_3$, Triton X-100, water | Standard addition | 0.2 µg $kg^{-1}$ | Aranda et al. (2009) |
| African palm, castor, palm, soybeans, and unknown oleaginous | Cd, Co, Cu, Mn, Ni, Pb, Ti, and Zn | ICP-MS | Microemulsion | $HNO_3$, Triton X-100, n-propanol | Aqueous standards in light mineral oil treated with n-propanol and Triton X-100 | $9.63.10^{-3}$–19.5 µg $L^{-1}$ | Amais et al. (2010) |

(continued)

**Table 6.1** (continued)

| Biodiesel | Element | Analytical technique | Sample preparation | Solutions mixture | Calibration | DL | References |
|---|---|---|---|---|---|---|---|
| Soybean, castor bean, animal fat, sunflower, and soybean + castor bean | Ca and Mg | F AAS | Microemulsion | Triton X-100, water + $HNO_3$, n-butanol or n-pentanol | Aqueous standards in mineral oil treated as samples | 0.04 and 0.005 µg $g^{-1}$ | de Jesus et al. (2010) |
| Soybean, animal fat, castor bean | Na and K | F AAS | Microemulsion | Triton X-100, n-butanol or n-pentanol, water + $HNO_3$ | Aqueous standards in base oil treated as samples | 0.1 and 0.06 µg $g^{-1}$ | de Jesus et al. (2008) |
| NR | Na, K, Ca, and Mg | F AAS | Microemulsion | n-propanol, $HNO_3$, water | Organometallic standard solutions | 0.004–0.1 µg $g^{-1}$ | Lyra et al. (2010) |
| Soybean, animal fat, sunflower, cotton | Ni and Cd | GF AAS | Microemulsion | Triton X-100, water + $HNO_3$ | Aqueous standards | ≤0.9 µg $L^{-1}$ | Lobo et al. (2011) |
| Soybean, canola, cotton, peanut, soybean + castor, barbados nut (Jatropha curcas L.) | As | GF AAS | Microemulsion | n-propanol, $HNO_3$ | 0.2% (v/v) $HNO_3$ aqueous solutions and organometallic standard solution | 0.3 mg $kg^{-1}$ | Vieira et al. (2009) |
| Fodder turnip, cottonseed, soybean, corn, and waste frying oils | Na and K | F AES | Microemulsion | HCl, water, n-propanol, | Aqueous standards in mineral oil treated as samples | <0.1 µg $g^{-1}$ | Chaves et al. (2008b) |
| Castor bean, soybean, soybean lees, cotton, | Cd, Pb, and Tl | GF AAS | Microemulsion | Water + $HNO_3$, n-propanol | Calibration against standard solutions prepared in aqueous medium and matrix-matching | 0.5–6 ng $g^{-1}$ | Silva et al. (2010) |

(continued)

**Table 6.1** (continued)

| Biodiesel | Element | Analytical technique | Sample preparation | Solutions mixture | Calibration | DL | References |
|---|---|---|---|---|---|---|---|
| animal fat, fodder turnip, and sunflower | | | | | calibration using microemulsion prepared with base oil | | |
| Soybean | Ca and Mg | F AAS | Emulsion breaking | Water, Triton X-114, $HNO_3$ | External calibration whit internal standardization (Co) | 0.014 and 0.004 μg $L^{-1}$ | Pereira et al. (2014) |
| Soybean | Cu, Mn, and Ni | ET AAS | Emulsion breaking | Triton X-100, water, $HNO_3$ | External calibration in aqueous medium | 0.053–0.30 | Pereira et al. (2013) |

*LD* detection limit, *RSD* relative standard deviation, *NR* not reported standards, low analyte stability in the diluted organic standards for calibration, variations in analyte concentration, and the necessity of high sensitivity determination techniques due to the dilution of the analyte in the solvents (Korn et al. 2007).

In a paper published by Lyra et al. (2010), the stability of Na, K, Ca, and Mg in biodiesel samples diluted by 1:10 (w/v) xylene was monitored by flame atomic absorption spectrometry. The results showed a significant decrease of the analyte absorbance (38% for Na, 66% for K, 22% for Ca, and 62% for Mg) in only 6 h. According to the authors, this decrease can be explained by the adsorption of the metal on the vessel walls, providing uncertain results which affect the precision of the method.

In addition, the use of organic solvents still presents a reasonable number of drawbacks when spectrometric techniques are employed. In ICP techniques, specific configurations of the torches, nebulizers, and modifications in the argon flux may be necessary to achieve satisfactory results, since the volatile nature of the organic solvents (Hardaway et al. 2004) and the possibility of destabilization or extinction of plasma (Korn et al. 2007) may jeopardize the measures. Organic solvents can produce flames very rich in fuel and unstable for FAAS. Regarding ETAAS, the difficulties in introducing the sample due to the low viscosity and surface tension and excessive spreading of the sample after its injection into the oven can be some drawbacks observed. By the way, interference caused by carbon-containing polyatomic species may occur in ICP-MS.

### *6.2.2 Emulsion and Microemulsion*

Emulsions are heterogeneous and kinetically stable systems, formed by mixing two immiscible or poorly miscible liquids, where one is dispersed in the form of droplets and the other is the matrix in which the droplets are dispersed, called the internal phase or the continuous phase (Burguera and Burguera 2012). The emulsion formed can be oil-in-water (O/W) type, if the oil droplets are dispersed in water, or water-in-oil (W/O) type if the water droplets are dispersed in oil (Viana et al. 2014).

Emulsions are colloidal systems, presenting a typically turbid due to the Tyndall effect with droplets ranging from 0.5 to 50 (Burguera and Burguera 2012). In addition, emulsions are thermodynamically unstable and non-spontaneous systems, which require mechanical or manual agitation of the solutions. The formation free energy of the emulsion is greater than zero ($\Delta G > 0$), and by this characteristic the phases are separated. Thus, when the agitation is interrupted, the droplets coalesce rapidly and the liquids are separated. Coalescence is often considered as the most important mechanism of destabilization of emulsions (Capek 2004).

The kinetic stability of the emulsions is easily improved by adding emulsifying agents (surfactants) to oil–water mixture. Some surfactants have amphiphilic characteristics, which their structure has a hydrophilic part (water soluble) and another part lipophilic (soluble in lipids). In other words, a typical single surfactant has the R-X structure, where R is a long hydrocarbon chain containing 8–18 C atoms, and X is a polar (or ionic) group. Depending on X, the surfactants may be classified as nonionic (e.g., Triton X-100, Triton X-114, Tween 80, Tween 60), anionic [e.g., sodium dodecyl sulfate (SDS), sodium dodecyl benzene sulfonate (DDBS)], cationic [e.g., hexadecyltrimethylammonium bromide (CTAB), cetylpyridinium chloride (CICP)], or amphoterics (e.g., N-alkyl and C-alkyl betaine and sultaine, alcohol aminophosphatidyl)] (Pelizzette and Pramauro 1985).

One of the major advantages of using emulsions for analytical purposes is the viability of using aqueous standard solutions in the calibration step. In this case, O/W emulsions are the most appropriate due to its low viscosity and present a long-time stability. However, in order to obtain a particular type of emulsion, the choice of the most suitable surfactant is challenging and requires a certain level of experience, given the wide variety of compounds with emulsifying characteristics. To support the choice of the ideal surfactant for a desired purpose, Griffin (1949) introduced the hydrophilic–lipophilic balance (HLB) concept and established a numerical ranks system for surfactants (on a scale of 1 to 20) according to their hydrophilicity or lipophilicity characteristics. This system is based on a fact that all surfactants have hydrophilic and lipophilic portions, and the balance between these portions varies from one to other compound. The surfactant becomes more hydrophilic when the HLB value increases, whereas, for lower HLB values, the compound is more lipophilic. Generally, in the preparation of emulsions, a surfactant with a suitable hydrophilic–lipophilic balance allows relatively high solubility between the immiscible phases. Surfactants which present a tendency to form O/W emulsions are hydrophilic and have HLB values ranging from 9.6 to 16.7 (Viana et al. 2014). In most cases, the surfactant is chosen by trial and error, although the ideal is learning to use the HLB values. However, these values apply only to nonionic surfactants (Burguera and Burguera 2012). In order to choose the most suitable surfactant for emulsion stabilization, a catalog provided by Sigma-Aldrich Co offers a variety of nonionic surfactants rated HLB (Merck KGaA 2017).

In relation to the concentration of surfactant used, it is recommended an amount from 10 to 20% (m/v) to the continuous phase (Thompson and Davidow 2013). But this amount can be best determined by means of an experimental design where different amounts of the oily sample and the surfactant are tested to give a stable emulsion.

Hoar and Schulman (1943) observed that the addition of alcohol to an emulsion (mixture of oil and a surfactant with water) could convert this initially turbid and unstable system in an optically clear and thermodynamically stable liquid. In 1959, the term microemulsion was used to designate this system (Schulman et al. 1959).

Generally, microemulsions are formed by the mixture of water and oil, stabilized by a surfactant combined with a cosurfactant, wherein the particle diameter is in order of micrometers (0.01–0.14 nm) (Silva et al. 2015; Burguera and Burguera 2012). These dispersed systems have low viscosity and are monophasic, isotropic, thermodynamically stable, optically transparent and presented very low interfacial tension. The microemulsions may incorporate large amounts of molecules which are generally insoluble in the continuous phase due to their high surface areas. So, the efficiency of the microemulsion is measured by its ability to solubilize molecules in their continuous phase (Khodakiya et al. 2012).

Although other components are present, due to their simplicity, microemulsions should be considered as ternary surfactant/oil/water systems (Burguera and Burguera 2012). The presence of surfactants can facilitate the emulsification and stabilization of the solutions by reducing interfacial tension, introduce double layer and/or solvation forces between the dispersed particles, and improve the interaction between the immiscible liquids (Pelizzette and Pramauro 1985).

The addition of a cosurfactant to the microemulsions preparation has the effect of neutralizing the repulsive effect between the polar heads of the surfactant allowing the formation of a membrane between the micelle and the continuous phase of the system, which prevents the significant increase of the viscosity of the microemulsion (Llorent-Martínez et al. 2011). In addition, the presence of alcohol can influence the solubility of the aqueous and oily phases, due to the partition of both phases (Dantas et al. 2003). The most used cosurfactants in microemulsions system are alcohols and glycols of small or medium molecular weight containing from two to ten carbons. Among the most common cosurfactants, 1-butanol, 2-butanol, 2-methyl-1-butanol, 1-pentanol, and 1-hexanol (Silva et al. 2015) are frequently used.

In general aspects, microemulsions are small-scale versions of the emulsions system and differ from the emulsions system due to its formation procedure, droplet size, physicochemical characteristics, and appearance (Table 6.2). In addition, the main emulsion types (O/W and W/O) involve the presence of two phases in equilibrium, while the structure of the microemulsion is classified into three distinct types: oil in water (O/W), water in oil (W/O), and bicontinuous structure (microemulsion in which microdomain oil and aqueous phase are randomly connected with approximately same volume) (Burguera and Burguera 2012). Although the amount of surfactant added in emulsions is very small, from 0.1% to 1.0% of the emulsion total mass, microemulsions are at least 10% of the total weight. These high levels of surfactants are essential due to the large increase of the interface area

**Table 6.2** Main characteristics presented by the emulsions and microemulsions solutions

| Parameters | Emulsion | Microemulsion |
|---|---|---|
| Formation | Mechanical or manual agitation | Spontaneous |
| Stable | Kinetically | Thermodynamically |
| Droplet size (μm) | 0.5–50 | 0.01–0.10 |
| Appearance | Opaque (cloudy) | Transparent |
| Surface area ($m^2$ $g^{-1}$) | Low (15) | High (200) |
| Interfacial tension | Low | Ultralow |
| Surfactant concentration | Low | High |
| Cosurfactant type | None | Short-chain alcohol |

Adapted from Burguera and Burguera (2012)

between the aqueous and oil phase (Khodakiya et al. 2012). On the other hand, a large sample dilution is necessary for the formation of the microemulsions, which leads to the development of procedures with lower limits of detection in relation to emulsions systems (Burguera and Burguera 2012).

In microemulsion systems, the goal is to obtain a critical combination between the components to achieve the maximum "solubilization" of the dispersed phase (Oliveira et al. 2004). When the components are grouped at a suitable ratio and when the interfacial tension is about $10^{-3}$ mN $m^{-1}$, the microemulsions are formed spontaneously and may present a similar behavior to aqueous solutions if the system is based on O/W microemulsion type (Mitchell and Ninham 1981). In this case, the phases diagram is useful to describe in which experimental condition it is possible to obtain microemulsions and also the transition regions between emulsions, separate phases, and O/W and W/O microemulsions (Oliveira et al. 2004). Usually, microemulsions are prepared by spontaneous emulsion method (phase titration method) and represented by pseudo-ternary phases diagram, where the aqueous phase, the oil phase, and surfactant/cosurfactant mixture are represented at the vertices of the triangle (Damasceno et al. 2011). The ideal formulation of the microemulsion corresponds to that required for the resulting system to have the desired physicochemical properties, such as viscosity, stability, dispersed phase specific area, technologically efficient application, and, of course, minimum cost (Oliveira et al. 2004).

Emulsified systems have been successfully applied for fuel samples in elemental determination carried out by spectrometric techniques, mainly because the emulsification system allows to high sensitivity levels without sample decomposition, as well as to use aqueous standards solution for calibration, and low costs. Moreover, analyte losses by volatilization or precipitation are eliminated, and contaminations are reduced because sample handling is minimal. Other significative aspects of the emulsification are the fact that the viscosity of the sample is closer to those obtained for aqueous solution and the analyte is easily introduced into the instrument, keeping its integrity as the original sample. Furthermore, the total organic carbon is reduced, and the system can be reconstituted by agitation without impairing the instrumental reproducibility (Burguera and Burguera 2012).

Raposo et al. (2014) developed a method for the determination of Na, K, and Ca in biodiesel emulsions from different sources by flame atomic emission spectrometry (FAES). Triton X-100 and X-114 were tested as surfactants, and ethanol, n-propanol, and n-butanol alcohols were evaluated as co-solvents. According to the authors, more stable and homogeneous emulsions were obtained when n-butanol was used as co-solvent, and the volume of 1.0 mL was sufficient to stabilize 1.0 g of sample. In order to dissolve suspended particles and convert organometallic species to inorganic species, as well as to improve the stability of the analytes in the emulsion, concentrated nitric acid was added to the biodiesel prior to the other components. Multivariate analysis through a central composite design (CCD) was applied in order to optimize the proportions of Triton X-100 and nitric acid. Recoveries in the 89–108% range demonstrated that the emulsification procedure was not influenced by matrix.

Aranda et al. (2012) evaluated the effect of the nitric acid concentration on the stability of biodiesel emulsions, and the results showed that more stable emulsions can be formed using 4:1 (m/v) of biodiesel:$HNO_3$. After addition of $HNO_3$, the aliquot of biodiesel was kept at rest for 15 min. Triton X-100 and isopropanol were added as surfactant and cosurfactant, respectively. The mixture was conducted to an ultrasonic bath for 15 min, and a desired volume of ultrapure water was added. Aliquots of 10 μL of the emulsions were directly injected into a graphite furnace atomic absorption spectrometer for the determination of Cd and Hg. The analyte addition method was used as a calibration strategy.

Amais et al. (2010) observed that light mineral oil microemulsions prepared in ethanol presented less stability compared to microemulsions prepared in n-propanol. According to the authors, this behavior can be explained based on the different polarity between ethanol and n-propanol. In other words, n-propanol more strongly interacts with the oil and the aqueous phase of the microemulsion because of the lower polarity than the ethanol. Based on this result, stable microemulsions of biodiesel from different sources (African palm, castor beans, palm, soybeans, and an unknown oleaginous) were prepared using 0.25 mL of 20% (v/v) $HNO_3$, 0.25 mL of Triton X-100, 0.5 mL of sample, and 4.0 mL of n-propanol. The mixture was homogenized using vortex stirrer for 2 min. Cadmium, Cu, Mn, Ni, Pb, Ti, and Zn were determined in the biodiesel microemulsions by inductively coupled plasma mass spectrometry (ICP-MS). The argon–oxygen mixture was added to the plasma as auxiliary gas to correct the matrix effects caused by the high carbon content in the biodiesel microemulsions. The accuracy of the proposed method was evaluated by addition and recovery tests and provided recoveries from 76.5 to 116.2% for all analytes except for Zn in castor oil biodiesel (65.0–76.2%). Recoveries below 86.6% were obtained for samples of palm biodiesel, probably due to matrix effects.

Oil-in-water (O/W) microemulsions, which are prepared with a large amount of water, are desirable due to reduced costs and facilitate the optimization of parameters for the analysis, as well as making possible the use of aqueous standard solutions, as already mentioned. Based on this perspective, de Jesus et al. (2008) attempted to prepare biodiesel microemulsions with large amounts of water for the determination of Na and K by flame atomic absorption spectrometry. However, an amount of 20% (m/m) of the hydrophobic phase (biodiesel) could not be stabilized when more than 8.0% (m/m) of water was used, and only emulsion formation was achieved. To avoid this drawback, the use of water-in-oil (W/O) microemulsions was chosen, using a maximum of 8.0% (m/m) water, and the use of aqueous standards for calibration was not feasible. With a pseudo-range phase diagram, the authors could determine the optimal proportions of Triton X-100 surfactant and cosurfactant (n-butanol or n-pentanol). This procedure was able to stabilize up to 90% (w/w) of the sample. Although the use of a maximum amount of sample in the microemulsion is desirable because it increases the sensitivity of the analysis, it also promotes increased viscosity and can reduce the efficiency of the nebulization. In this way, the researchers adopted the amount of 20% (m/m) of sample to continue the study. Finally, the 57.6% (m/m) of n-pentanol or n-butanol, 20% (m/m) of biodiesel, 14.4% (m/m) of Triton X-100, and 8.0% (m/m) of water were adopted for microemulsion preparation. According to the authors, the K and Na concentration were stable for 3 and 5 days, respectively. In other words, there were no significant changes in the analyte signals during the time required to perform the analysis, and the proposed sample preparation procedure was considered satisfactory for routine analyzes.

Although surfactants are generally considered a component necessary for the formation of the microemulsion, the literature has shown that even in the absence of traditional surfactants, microemulsions can also be formed (Smith et al. 1977; Ni and Hou 2008; Song et al. 2010; Klossek et al. 2012; Xu et al. 2013). These microemulsions, called surfactant-free microemulsions (SFMEs), constitute a ternary system of two immiscible liquids and a solvent partially or completely miscible with each immiscible liquid, called amphi-solvent (Xu et al. 2013). Among the amphi-solvents used to form SFMEs, those contain short-chain alcohols (ethanol, propanol, and butanol), N, N-dimethyl formamide (DMF), acetone, and tetrahydrofurfuryl alcohol (THFA) (Hou and Xu 2016) have been satisfactorily applied. Partly, the structures and properties of SFMEs are similar to those observed to traditional surfactant-based microemulsions (SBMEs). However, emulsions without surfactants are turbid and thermodynamically unstable and can remain well dispersed for months, but the separation of the phases will inevitably occur (Sakai 2008). A method for determining Na, K, Ca, and Mg in surfactant-free microemulsion of biodiesel by flame atomic absorption spectrometry was developed by Lyra et al. (2010). The optimum composition of the microemulsion was 10% (v/v) biodiesel, 75% (v/v) n-propanol, 1.0% (v/v) concentrated nitric acid, and 14% (v/v) aqueous solution formed by 0.2% (v/v) nitric acid and 0.5% (v/v) ionization suppressor. According to the authors, this mixture resulted in homogeneous and transparent microemulsions. The stability of the analytes in the microemulsions was monitored, and it was shown that the absorbances remained stable for a period of 15 days. The recovery obtained for the analytes ranged from 89 to 101%, which indicates that there was no influence of the matrix on the preparation of the microemulsions.

### *6.2.3 Extraction Induced by Emulsion Breaking*

The procedure based on extraction induced by emulsion breaking is a sample preparation approach initially proposed by Cassella et al. (2010), which consists of the extraction of the selected elements from the oily samples followed by the determination by spectrometric techniques. Since then, this method has been successfully applied in the elemental analysis of different types of oily samples, such as diesel oil (Cassella et al. 2012), gasoline (Vicentino and Cassella 2017), lubricating oil (Caldas et al. 2013), edible oils (Bakircioglu et al. 2014), and also in biodiesel samples (Pereira et al. 2014, 2013). The method is based on the formation and breaking of the emulsions prepared by the mixture of oily sample and acidified aqueous phase containing the surfactant. During the emulsion breaking process, which may be induced by heating or centrifugation, the analytes are transferred to the aqueous phase which is separated from the oil.

One of the most important aspects of the extraction induced by emulsion breaking is to establish the appropriate operating conditions for quantitative transfer of the selected analyte in the organic phase to the aqueous phase (Caldas et al. 2013). In this context, the literature has presented studies showing the influence of the type and concentration of both surfactant and acid, the time required for emulsion breaking, and the efficiency of the extraction of the analyte.

In a work carried out to determine Cu by graphite furnace atomic absorption spectrometry, Caldas et al. (2013) found that the type and concentration of surfactants (Triton X-100, Triton X-114, and Tween 20) had no influence on extraction efficiency of the metal from mineral oil. According to the authors, this behavior was expected, since the surfactant added to the extractive solution should only act as emulsifying agent, in order to allow a more effective contact between the organic sample and the aqueous extractive phase. On the other hand, both concentration and type of surfactant used exerted a strong influence on heating times required for emulsions breaking. The extraction time increased, mainly due to a greater stability of the emulsions caused by the increase of the surfactant concentration. Thus, the use of Triton X-114 is more suitable for the proposed method, since it led to the formation of less stable emulsions, with less heating time required for emulsion breaking than the other surfactants tested. In this same work, the authors evaluated the effect of the concentration and type of the acid (HCl, $HNO_3$, and the $HNO_3$/HCl mixture) used as extractive solution. Lower analytical signals were observed for extracts obtained in free acid emulsions, which, according to the authors, reinforces the idea that the acid interferes with the extraction process. At concentrations higher than 10% (v/v), no significant variation in the Cu analytical signal was observed, but the extraction efficiency is higher when nitric acid is used, possibly due to its greater oxidizing character in relation to hydrochloric acid. The researchers also found that the heating time needed to achieve emulsion breakdown decreases with increasing acid concentration, with more significant effect using $HNO_3$.

In other work, the extraction induced by emulsion breaking procedure was used to biodiesel samples to determine Cu, Mn, and Ni by electrothermal atomic absorption spectrometry (Pereira et al. 2013). In this procedure, the authors evaluated some parameters that could influence the performance of the method. According to authors, the concentrations of $HNO_3$, Triton X-100, and extraction time were the variables that most influenced the development of the procedure. The recoveries varied from 89 to 109%, which indicated that the metals were extracted quantitatively from the biodiesel using the proposed procedure, and showed no matrix interference by ET AAS.

Pereira et al. (2014) reported the determination of Ca and Mg in biodiesel by FS-FAAS after extraction induced by emulsion breaking. After establishing the suitable conditions for the procedure, the quantitative extraction of the analytes was obtained from 20 mL of biodiesel and 4.0 mL of a solution containing 2.1 mol $L^{-1}$ of $HNO_3$ and 7.5% (m/v) of Triton X-114. Only after the emulsions formation, the flask was heated at 90 °C in a water bath until emulsion breaking was observed, which took approximately 10 min. Although the analytes are extracted into the aqueous phase in this sample preparation approach, in this study it was reported that the presence of surfactant in the extracts caused random errors, possibly because it was affected by the nebulization process and the transport efficiency of the solution to flame, making the external calibration strategy unfeasible. In order to overcome this problem, the internal standardization with Co was adopted as a calibration strategy.

### *6.2.4 Dispersive Liquid–Liquid Extraction: Fundamentals and Applications*

Despite the significant advances in analytical instruments and approaches developed in recent years, the sample pretreatment procedures still are considered the most important step in an analytical process, since it is used for the isolation and concentration of components of interest or target analytes from different sample matrices and ensures the validity of the results obtained (Assadi et al. 2012). Based on this, great effort has been made by researches to improve the performance of sample preparation procedures, following the currents trends highly focused on improving the quality of analytical results, lower reagents consumption, high selectivity and recoveries using environmental-friendly methods, and introduction of new technological developments involving miniaturization, simplification, and automation (Al-Saidi and Emara 2014). Among the procedures previously established in the literature, the liquid–liquid extraction (LLE) still is one of the most traditional and well known for elemental determination in different samples, in which it is based on the transfer of analyte from the aqueous phase to a water immiscible solvent. Nevertheless, some shortcomings such as use of large sample volumes and toxic organic solvents, emulsion formation that obstructs automation of the technique and generation of large amount pollutants make ELL a tedious procedure with long extraction times, high cost due to the use of solvents of high purity and environmentally unviable (Rezaee et al. 2010). Therefore, the development of new sample preparation procedures that provide accurate and precise results with reasonable quantification limits coupled with greater efficiency extraction, less time consuming, and smaller amounts of organic solvents are extremely desirable. Miniaturization and improvement of LLE have been started by several researches since mid- to late-1990s, through strategies using a drastic reduction of the extractant phase volume. As example, liquid-phase microextraction (LPME) and single-drop microextraction (SDME) emerged as alternative miniaturized sample preparation approach using microliter volume of the solvent to extract analytes from the aqueous sample. These methods are inexpensive and since very little solvent is used, there is minimal exposure to toxic organic solvents. However, some disadvantages such as breaking of the organic drop by fast stirring, formation of air bubble, and time-consuming extraction propelled some modifications in order to improve the stability and reliability of LPME methods. Among them, hollow fiber liquid-phase microextraction (HF-LPME), liquid-phase microextraction with solidified floating organic droplet (LPME-SFO), cloud point extraction (CPE), homogeneous liquid-liquid extraction (HLLE) and liquid-liquid-liquid microextraction (LLLME) deserve attention. These procedures have numerous positive features including: simplicity of operation, relatively low costs of instrumentation, versatility, and short time of extraction. Although these sample preparations have gained their place in the field of analytical sciences, it faces their own constraints, which makes the development of sample preparation techniques a continuous process (Assadi et al. 2012; Rezaee et al. 2010; Xiao-Huan et al. 2009; Yan and Wang 2013).

One of the techniques that have attracted special attention is dispersive liquid–liquid microextraction (DLLME) introduced by Rezaee et al. (2006) for the preconcentration of organic and inorganic analytes from aqueous matrices (Rezaee et al. 2006). As the name suggests, DLLME is equivalent to a miniaturized type of LLE that is based on the formation of a cloudy solution (highly turbulent) when a solvent mixture, composed of a small amount of the extractant solvent and a larger amount of the disperser solvent, is injected into the sample solution by syringe. The disperser solvent is miscible in both aqueous and organic phases. This turbulent gives rise to the formation of small droplets which are dispersed throughout the aqueous sample increasing the surface area between the extraction solvent and the aqueous sample. After centrifugation, the dispersed fine particles of extraction phase sediment on the bottom of the bottle are taken with a microsyringe and subsequently analyzed. In addition, the equilibrium state is achieved quickly resulting in extraction time is very short, which characterizes the main advantage of this method compared with those of the other techniques (Maya et al. 2014). In light of the merits of DLLME, such as simplicity and efficiency of the extraction process, low cost, high enrichment factor (EF) due to the high ratio of the aqueous phase (sample) and extraction solvent, lack of negative environmental impact, DLLME procedure seems to meet the requirements of green chemistry and to be compatible with most of the analytical techniques, such as high-performance liquid chromatography (HPLC), gas chromatography (GC), electrothermal atomic absorption spectrometry (ETAAS), and flame atomic absorption spectrometry (FAAS).

It should be noted that the applicability of DLLME is mainly based upon the distribution coefficient ($K$) defined as the ratio between the analyte concentration in extraction solvent and sample solution. Thus, analytes with high or moderate lipophilic property ($K > 500$) are favored in the extraction process, unlike neutral analytes with high hydrophilic properties. However, for the alkaline or acidic analytes, distribution coefficient could be increased by controlling the pH of sample solution, resulting in analytes in their nonionic state. Some parameters such as enrichment factor (EF) and extraction recovery (ER) indicate the performance and efficiency of DLLME in different sample media. In conventional DLLME, EF is defined as the ratio between the analyte concentration in the sedimented phase ($C_{sed}$) and the initial concentration of analyte ($C_0$) within the sample (Eq. 6.1).

$$\mathrm{EF} = C_{\mathrm{set}}/C_0 \tag{6.1}$$

where $C_{set}$ can be obtained from a suitable calibration or by conventional LLE. The extraction recovery is defined as the percentage of the total analyte amount ($n_0$) extracted to the sedimented phase ($n_{sed}$) and represented by the following equations:

$$\mathrm{ER}(\%) = n_{\mathrm{sed}}/n_0 \times 100 = C_{\mathrm{sed}}/C_0 \times V_{\mathrm{sed}}/V_{\mathrm{aq}} \times 100 \tag{6.2}$$

$$\mathrm{ER}(\%) = V_{\mathrm{sed}}/V_{\mathrm{aq}} \times \mathrm{EF} \times 100 \tag{6.3}$$

where $n_0$ and $n_{sed}$ represent the total and sedimented amount of the analyte, whereas $V_{sed}$ and $V_{aq}$ are the volumes of sedimented phase and sample solution, respectively. The extraction efficiency for the target analyte by DLLME is influenced by many factors: kind and volume of extraction and disperser solvent, ionic strength, pH, extraction time, and salt addition. It is very important to optimize these in order to obtain a higher enrichment factor and extraction recovery (Zgola-Grzeskowiak and Grzeskowiak 2011). Several strategies have been employed for optimization of these parameters in DLLME, among them one-variable-at-a-time optimization is frequently used because of its simplicity. However, some other chemometrics strategies such as central composite, factorial design, and others have been used to date.

#### 6.2.4.1 Effects of Extraction and Disperser Solvent in DLLME

The choice of an appropriate extraction solvent is one of the most important parameters for the DLLME process. Solvents differ in their extraction capabilities depending on their own and the solute's chemical structure. High distribution and partition coefficient, good selectivity toward solution, and little or no miscibility with aqueous solution are the main desirable properties of the solvents used in extraction procedures. In addition, other factors such as boiling point, interfacial tension, density, viscosity, toxicity, stability, and cost also affect solvent selection. As already mentioned above, extraction solvents commonly utilized in DLLME are selected for having different density of water, extraction capability of target compounds, low solubility in water, and miscibility with disperser solvent, besides to being compatible with the majority of the instrumental techniques. For this, different classes of extractants are being used in DLLME, like solvents with densities higher than water (density > 1), solvents with densities less than water, and more recently, ionic liquids and supramolecular solvents. Most of the high-density extraction solvents are halogenated solvents such as chlorobenzene, chloroform, carbon tetrachloride, tetrachloroethylene, 1,2-dibromoethane or nonhalogenated solvents such as carbon disulfide and nitrobenzene. Although the high-density extraction solvents are toxic to humans and the environment, the use of lower density solvents makes the analysis relatively tedious and relatively complex, since the collected phase for further determination is on the upper surface of the sample solution as a microdrop or a thin film after centrifugation step. In this case, to overcome this disadvantage and avoid using different techniques to remove the less dense solvent, as capillary tube (Farajzadeh et al. 2010) or specialized extraction vessel (Farajzadeh et al. 2009), an auxiliary solvent can be used to increase mixture's density favoring the phase separation step by centrifugation (Al-Saidi et al. 2015; Rezaee et al. 2010).

Because the high toxicity of extraction solvents commonly used in DLLME and its limitation in extracting various analytes with different polarities, great attention has been given to the search for new extraction solvents that are polar and low density. For example, room temperature ionic liquids (ILs) have been used successfully as extraction solvent in DLLME due to their unique physicochemical properties which depend on the nature and size of their cationic and anionic constituents. These include negligible vapor pressure, nonflammability, tunable viscosity, and miscibility with water and organic solvents (Al-Saidi and Emara 2014; Assadi et al. 2012; Rezaee et al. 2010; Zgola-Grzeskowiak and Grzeskowiak 2011). ILs are considered a class of liquids entirely comprised of ions, which usually exist as liquids well below room temperature. Among frequently used ILs, 1-hexyl-3-methylimidazoliumhexafluorophosphate ([Hmim][$PF_6$]), 1-octyl-3-methylimidazolium bis(trifluoromethylsulfonyl)imide, [Omim][$TF_2N$], and 1-hexyl-3-methylimmidazolium bis(trifluormethylsulfonyl)imid, ([Hmim] [$TF_2N$]) (Gharehbaghi et al. 2009; Molaakbari et al. 2011; Yousefi and Ahmadi 2011) are successfully employed as extraction solventes.

Like ILs, supramolecular solvents have been proposed in DLLME as an alternative way to alleviate the impact of organic solvents. Extraction using supramolecular solvent occurs due to a lyophilized colloid, composed of a set of separate droplets and constituted by aggregates of surfactants, comprising the aqueous phase. Its agglomeration property can be modified by altering the nature of the macromolecule or the coacervating agent (coagulant). Depending on the composition of the supramolecular solvent, analyte extraction may be based on different non-covalent interactions, e.g., hydrophobic interactions with the hydrocarbon chains or hydrogen bonds with the carboxylic and carboxylate groups. A mixture of decanoic acid and tetrabutylammonium as coacervating agent has been proposed as supramolecular solvent in several papers (López-Jimenez et al. 2008; Moral et al. 2009; Jafarvand and Shemirani 2011a, b, c). The most significant contribution of supramolecular solvent in DLLME process is related to improved extraction capability for analytes in a wide range of polarities/charges rather than only hydrophobic analytes. It should be noted that the use of decanoic acid was also highlighted since it meets the criteria of green chemistry.

Similarly, to the solvent extractor, special attention should be paid to the selection of the disperser solvent. It should be miscible in both the extraction solvent and the aqueous sample, thus enabling the extraction solvent to be dispersed as fine particle in aqueous phase to form a cloudy solution (water/disperser solvent/ extraction solvent). There are different groups of disperser solvents based on their natures and dispersion techniques. Among them, the most usually employed are acetone, methanol, ethanol, tetrahydrofuran, and acetonitrile because of their high miscibility in both phases (organic and aqueous).

Besides the choice of extraction solvent, the exact determination of its volume has a great effect on the enrichment factor. Thus, with the increase of the extraction solvent volume, the final organic phase after centrifugation step is increased, resulting in a lower concentration of the target analyte in organic phase and lower EF. Therefore, the determination of the optimum volume of extractive solvent (in general, 5–100 μL) is important to ensure both high EF and enough volume of the sedimented phase for the subsequent analysis after centrifugation step. Already disperser solvent volume directly affects the formation of cloudy solution, the dispersion degree of the extraction solvent in aqueous phase, and consequently the extraction efficiency (Hashemi et al. 2017; Xiao-Huan et al. 2009; Zgola-Grzeskowiak and Grzeskowiak 2011). In addition, as disperser solvent plays an important role in decreasing the interfacial tension between water and extracting solvent, low volumes not disperse the extraction solvent properly, decreasing the solution turbulence significantly. Instead of this, high volumes increase the solubility of analytes in water phase, making the extraction process incomplete.

#### 6.2.4.2 Effects of Extraction Time in DLLME

Mass transfer of analytes from the sample solution to the extraction phase is a time-dependent process. In DLLME, extraction time is defined as the interval time between the injection of extraction mixture (extraction solvent:disperser solvent) and the starting of the centrifugation. In general, the equilibrium state in this extraction is achieved very quickly, since the surface area between the extraction solvent and the aqueous phase is infinitely large after the formation of cloudy solution. Thus, the majority of studies of DLLME have shown that the extraction time is very short, being a remarkable advantage of the DLLME technique (Assadi et al. 2012; Xiao-Huan et al. 2009).

#### 6.2.4.3 pH

Although the extractions of most compounds are independent of pH, in some cases the pH of the sample solution plays an important role in complex formation or the derivatization reaction. Organic compounds can be extracted from aqueous phase in nonionized form. However, for ionizable compounds the adjustment of the solution pH favors the formation of the compound in its non-ionizable state (Assadi et al. 2012; Zgola-Grzeskowiak and Grzeskowiak 2011).

#### 6.2.4.4 Effects of the Ionic Strength in DLLME

Salt addition is frequently used in extraction process to adjust the ionic strength, improve the extraction efficiency, and reduce the detection limit. However, in DLLME two opposite effects were observed when the ionic strength of solution increases. For analytes hydrophilic, the salt addition usually increases the enrichment factor by increasing the extraction efficiency. On the other hand, by increasing the amount of salts (e.g., NaCl, $KNO_3$, KCl), the volume of the sedimented phase would increase slightly due to the reduction in solubility of the extraction solvent in the sample, therefore reducing the enrichment factor. In addition, the phase separation is impaired when high concentration of salts is used (Assadi et al. 2012; Zgola-Grzeskowiak and Grzeskowiak 2011).

#### 6.2.4.5 Application of DLLME in Elemental Determination

DLLME has been successfully and conveniently applied to the extraction/preconcentration of a wide range of organic compounds and metal ions, mainly from water samples. After pesticides, metals are the second most common analytes that are extracted by DLLME. The toxicity of trace metals in environmental, food, clinical, and biology samples is still an area of growing interest due to the fact that the toxicity of many metallic species depends on their form present in samples. Atomic absorption spectrometry is a powerful technique for the determination of trace metals in complex matrices, and it has been widely combined with DLLME since such technique only requires small amounts of sample. In addition, techniques such as ICP-OES and UV-Vis spectrophotometry also have been explored in DLLME studies. The versatility of DLLME combined with spectrophotometric techniques for extraction of metals from various matrixes is depicted in Table 6.3.

As can be seen in Table 6.3, water samples (mainly tap, river, lake, sea, and wastewater) have been the matrix field in which DLLME has found its major application due to the ease of execution as well as the weak purification ability. These samples are considered simple matrices and can be directly performed by the DLLME procedure or in some cases only require centrifugation step (Al-Saidi and Emara 2014). However, there is still a deficiency in the development of analytical procedures of sample preparation efficiently able to check the chemical quality of samples of great relevance in the international scenario, such as biodiesel. Although Hashemi et al. (2017) reported recently the advances of the liquid-phase microextraction techniques describing different approaches related to DLLME for analysis of environmental pollutants, little has been described regarding the use of DLLME as a sample preparation procedure for oil samples. One of the first papers reporting the use of DLLME in oil samples was described by López-García et al. (2014). Based on this, an analytical method of dispersive liquid–liquid microextraction for removal of Cd and Pb ions from edible oils was successfully proposed and carried out in a reverse mode, using low volume of an acidic solution in the presence of isopropyl alcohol. Unlike conventional DLLME, in the reverse dispersive liquid–liquid microextraction the analytes are extracted from an organic phase (oil samples) into the aqueous phase. Under the optimized experimental conditions of some parameters that directly affect the performance of DLLME, high preconcentration for both analytes was achieved. The enrichment factors obtained experimentally were of 140 and 146 for cadmium and lead, respectively, whereas

**Table 6.3** Application of DLLME combined with different spectrometric techniques in the determination of ions

| Analyte | Matrix | Extraction method | Analytical technique | EF | LD | References |
|---|---|---|---|---|---|---|
| Pd | Soil, water, and food | DLLME | FAAS | 20.7 | 8 | Pouyan et al. (2016) |
| Cd and Pb | Honey | DLLME | FAAS | – | 20.0 ng $g^{-1}$ Cd<br>140 ng $g^{-1}$ Pb | Rosa et al. (2015) |
| Cd and Pb | Edible oils | Reverse DLLME | ETAAS | 140 | 0.6 ng $kg^{-1}$ Cd<br>10.0 ng $kg^{-1}$ Pb | López-García et al. (2014) |
| Hg | Urine samples | Vórtex-assisted IL-DLLME | Gold nanoparticle-modified screen-printed electrodes | – | 0.5–1.5 | Fernández et al. (2016) |
| AgNPs | Water samples | IL-DLLME | ICP-MS | >90 | 0.01 | Chen et al. (2016) |
| Cd and Pb | Soft drinks | DLLME | GF AAS | ±200 | 0.006 ng $L^{-1}$ Cd 0.072 ng $L^{-1}$ Pb | Mandlate et al. (2017) |
| Cr (VI) | Water samples | DLLME | Uv-Vis | – | 7.48 | Ahmad et al. (2016) |
| Pd | Water and road dust sample | UASMDLLME-SFO | FAAS | 93 | 0.63 | Ezoddin et al. (2015) |
| As (III) Se (IV) Te (IV) | Certified reference materials and real water samples | DLLME | ETV-ICP-MS | As (III) 61 Se (IV) 107 Te (IV) 95 | 2.5 ng $L^{-1}$ As (III) 8.6 ng $L^{-1}$ Se (IV) 0.56 ng $L^{-1}$ Te (IV) | Liu et al. (2015) |
| Bi | Certified reference materials and water and hair | DLLME | UV-Vis | – | 0.54 | Al-Saidi et al. (2015) |
| Cr, Cu, Mn, Ni and Zn | CRM estuarine water | DLLME | LIBS | Cr: 4.8 Cu: 4.0 Mn: 5.5 Ni: 4.7 Zn: 5.5 | 41 µg $kg^{-1}$ Cr 23 µg $Kg^{-1}$ Cu 49 µg $Kg^{-1}$ Mn 107 µg $Kg^{-1}$ Ni 18 µg $Kg^{-1}$ Zn | Aguirre et al. (2015) |

(continued)

**Table 6.3** (continued)

| Analyte | Matrix | Extraction method | Analytical technique | EF | LD | References |
|---|---|---|---|---|---|---|
| Cu | Cereals and vegetable food samples | DLLME | FAAS | 55 | 0.05 ng $mL^{-1}$ | Shrivas and Jaiswal (2013) |
| Hg | Tap and mineral water | VALLME | UV-Vis | – | 0.8 ng $mL^{-1}$ | Martinis and Wuilloud (2016) |
| As (III) | Water samples | DLLME-SFO | GFAAS | 1520 | 2.5 ng $L^{-1}$ | Shamsipur et al. (2014) |
| Co | Food, environmental, and water samples | DLLME | FAAS | 8 | 12.4 ng $mL^{-1}$ | Ojeda et al. (2012) |
| Zn and Cd | Biological and water sample | LL-DLLME | FAAS | Zn: 10.30 Cd: 13.4 | 0.3 ng $mL^{-1}$ Zn 0.4 ng $mL^{-1}$ Cd | Mohammadi et al. (2012) |
| Tl (III) | Water sample | DLLME-SFO | FAAS | 17 | 90 ng $mL^{-1}$ | Mahmoud et al. (2014) |
| Cd | Human biological samples | DLLME-TSIL | ETAAS | 10.4 | 5 ng $L^{-1}$ | Shirkhanloo et al. (2016) |
| AuNPs | Water samples | SA-DLLME | ETV-ICP-MS | 152 | 2.2 ng $L^{-1}$ | Liu et al. (2016) |
| Cu | Drinking water and serum of adolescent female hepatitis C | IL-µE-DLLME | FAAS | 70 | 0.132 µg $L^{-1}$ | Arain et al. (2016) |
| Cu (I) and Cu (II) | Mineral water | GA-DLPME | UV-vis | 122 | 0.07 µg $mL^{-1}$ | Akhond et al. (2016) |

(continued)

**Table 6.3** (continued)

| Analyte | Matrix | Extraction method | Analytical technique | EF | LD | References |
|---|---|---|---|---|---|---|
| Cr(VI) | Human biological samples | CP-DILLME | ETAAS | 25.2 | 5.4 ng $L^{-1}$ | Shirkhanloo et al. (2016) |
| Au and Pd | CRM | DLLME | FAAS | Au: 94 Pd: 113 | 0.4 μg $L^{-1}$ Au 0.6 μg $L^{-1}$ Pd | Ozdemir et al. (2014) |
| Hg | Water samples | In situ IL-DLLME | SPCnAuEs | 25 | 0.2 μg $L^{-1}$ | Fernández et al. (2015) |

*AgNPs* Silver nanoparticles; *IL* Ionic liquid; *UASMDLLME-SFO* Ultrasonic-assisted supramolecular dispersive liquid–liquid microextraction based on solidification of a floating organic droplet; *ETV-ICP-MS* Electrothermal vaporization inductively coupled plasma mass spectrometry; *CRM* Certified reference materials; *VALLME* Vortex-assisted liquid–liquid microextraction; *DLLME-SFO* Dispersive liquid–liquid microextraction based on the solidification of floating organic drop; *LL-DLLME* Ligandless dispersive liquid–liquid microextraction; *DLLME-TSIL* Dispersive liquid–liquid microextraction based on a task-specific ionic liquid; *SA-DLLME* Surfactant-assisted dispersive liquid–liquid microextraction; *IL-μE-DLLME* Dispersive liquid–liquid microextraction procedure based on ionic liquid-assisted microemulsion; *GA-DLPME* Gas-assisted dispersive liquid-phase microextraction; *CP-DILLME* Cloud point-assisted dispersive ionic liquid–liquid microextraction; *In situ- IL-DLLME* in situ ionic liquid formation dispersive liquid–liquid microextraction; *SPCnAuEs* Gold nanostructured screen-printed carbon electrodes the detection limit, calculated on the basis of the criterion of three times the standard error from the calibration graphs, was equal 0.6 ng $kg^{-1}$ for Cd and 10 ng $kg^{-1}$ for Pb. In addition, it has also been demonstrated through studies that reverse DLLME procedure does not require sample mineralization and that the levels extremely low of Cd and Pb were only possible to measure due to the high preconcentration factor achieved in the DLLME process. Similarly, in 2015 López-García and coworkers reported the use of reverse dispersive liquid–liquid microextraction to extract water-soluble arsenic, such as inorganic arsenic, methylarsonate, dimethylarsinate, and arsenobetaine from the samples marketed as edible oils into a slightly acidic aqueous medium. The total arsenic content was measured in the extracts by ETAAS using palladium as the chemical modifier. Under optimized conditions, very low arsenic concentrations could be measured in the aqueous extracts due to the high preconcentration effect inherent in the reverse DLLME process, reaching the detection limit of 0.03 ng As per gram of oil. In addition, to verify the reliability of the proposed method, addition and recovery studies were performed and the percentage of arsenic recovery in the extracts was 94–105%, using 300 measures in total (López-García et al. 2015).

As can be seen in this chapter, the most adequate sample preparation procedures for oily sample are still considered a great challenge due to the matrix complexity. However, several improvements have been made in recent years following trends in modern analytical chemistry, such as the development of analytical methods of sample preparation that involves extraction step without use or with lower volumes of organic solvents reducing the negative impact on the environment and human health. By the way, although all of these techniques are characterized by a number of advantages, it has also several drawbacks and limitations. Thus, the development of new methodologies for biodiesel still is required taking into account green chemistry recommendations and the desirable extraction parameters, such as enrichment factors, low detection limit, and high frequency of sampling.

## References

ABNT NBR 15553:2008 (2008) Fat oil and derivatives. Fatty Acid Methyl Esters. Determination of the Calcium, Magnesium, Sodium, Phosphorus and Potassium Contents by Inductively Coupled Plasma Optical Emission Spectrometry (ICP OES), São Paulo

ABNT NBR 15556:2008 (2008) Fat Oil and Derivatives. Fatty Acid Methyl Esters. Determination of the Sodium, Potassium, Calcium and Magnesium Contents by Atomic Absorption Spectrometry, São Paulo

Aguirre MA, Selva EJ, Hidalgo M, Canals A (2015) Dispersive liquid–liquid microextraction for metals enrichment: a useful strategy for improving sensitivity of laser-induced breakdown spectroscopy in liquid samples analysis. Talanta 131:348–353

Ahmad W, Bashammakh AS, Al-Sibaai AA, Alwael H, El-Shahawi MS (2016) Trace determination of Cr(III) and Cr(VI) species in water samples via dispersive liquid-liquid microextraction and microvolume UV–Vis spectrometry. Thermodynamics, speciation study. J Mol Liq 224:1242–1248

Akhond M, Absalan G, Pourshamsi T, Ramezani AM (2016) Gas—assisted dispersive liquid-phase microextraction using ionic liquid as extracting solvent for spectrophotometric speciation of copper. Talanta 154:461–466

Al-Saidi HM, Emara AAA (2014) The recent developments in dispersive liquid–liquid microextraction for preconcentration and determination of inorganic analytes. J Saudi Chem Soc 18:745–761

Al-Saidi HM, Abdel-Fadeel MA, El-Sonbati AZ, El-Bindary AA (2015) Determination of bismuth in different samples by dispersive liquid–liquid microextraction coupled with microvolume β-correction spectrophotometry. J Mol Liq 212:635–640

Amais RS, Garcia EE, Monteiro MR, Nogueira ARA, Nóbrega JA (2010) Direct analysis of biodiesel microemulsions using an inductively coupled plasma mass spectrometry. Microchem J 96:146–150

Amais RS, Donati GL, Schiavo D, Nóbrega JA (2013) A simple dilute-and-shoot procedure for Si determination in diesel and biodiesel by microwave-induced plasma optical emission spectrometry. Microchem J 106:318–322

Arain SA, Kazi TG, Afridi HI, Arain MS, Panhwar AH, Khan N, Baig JA, Shah F (2016) A new dispersive liquid–liquid microextraction using ionic liquid based microemulsion coupled with cloud point extraction for determination of copper in serum and water samples. Ecotoxicol Environ Saf 126:186–192

Aranda PR, Pacheco PH, Olsina RA, Martinez LD, Gil RA (2009) Total and inorganic mercury determination in biodiesel by emulsion sample introduction and FI-CV-AFS after multivariate optimization. J Anal At Spectrom 24:1441–1445

Aranda PR, Gásquez JA, Olsina RA, Martinez LD, Gil RA (2012) Method development for Cd and Hg determination in biodiesel by electrothermal atomic absorption spectrometry with emulsion sample introduction. Talanta 101:353–356

Assadi Y, Farajzadeh M, Bidari A (2012) Dispersive Liquid–Liquid microextraction. In: Comprehensive sampling and sample preparation. Elsevier, pp 181–212

ASTM D6751-10 (2010) Standard specification for biodiesel fuel blend stock (B100) for middle distillate fuels, West Conshohocken, PA, USA

Avila Orozco FD, Kovachev N, Aguirre Pastor MÁ, Domini CE, Fernández Band BS, Canals Hernández A (2014) Analysis of metals and phosphorus in biodiesel B100 from different feedstock using a Flow Blurring® multinebulizer in inductively coupled plasma-optical emission spectrometry. Anal Chim Acta 827:15–21

Bader NR, Zimmermann B (2012) Sample preparation for atomic spectroscopic analysis: an overview. Adv Appl Sci Res 1733–1737

Bakircioglu D, Topraksever N, Kurtulus YB (2014) Determination of zinc in edible oils by flow injection FAAS after extraction induced by emulsion breaking procedure. Food Chem 151:219–224

Barros AI, de Oliveira AP, de Magalhães MRL, Villa RD (2012) Determination of sodium and potassium in biodiesel by flame atomic emission spectrometry, with dissolution in ethanol as a single sample preparation step. Fuel 93:381–384

BS EN 14108/2003 (2003) Fat and oil derivatives. Fatty Acid Methyl Esters (FAME). Determination of sodium content by atomic absorption spectrometry

BS EN 14109/2003 (2003) Fat and oil derivatives. Fatty Acid Methyl Esters (FAME). Determination of potassium content by atomic absorption spectrometry

Burguera JL, Burguera M (2012) Analytical applications of emulsions and microemulsions. Talanta 96:11–20

Caldas LFS, Brum DM, de Paula CER, Cassella RJ (2013) Application of the extraction induced by emulsion breaking for the determination of Cu, Fe and Mn in used lubricating oils by flame atomic absorption spectrometry. Talanta 110:21–27

Capek I (2004) Degradation of kinetically-stable o/w emulsions. Adv Colloid Interface Sci 107:125–155

Cassella RJ, Brum DM, de Paula CER, Lima CF (2010) Extraction induced by emulsion breaking: a novel strategy for the trace metals determination in diesel oil samples by electrothermal atomic absorption spectrometry. J Anal At Spectrom 25:1704–1711

Cassella RJ, Brum DM, Robaina NF et al (2012) Extraction induced by emulsion breaking for metals determination in diesel oil by ICP-MS. J Anal Spectrom 27:364–370

Chaves ES, Lepri FG, Silva JSA, de Quadros DPC, Saint'Pierre TD, Curtius AJ (2008a) Determination of Co, Cu, Fe, Mn, Ni and V in diesel and biodiesel samples by ETV-ICP-MS. J Environ Monit 10:1211–1216

Chaves ES, Saint' Pierre TD, dos Santos EJ, Tormen L, Bascuñan VLAF, Curtius AJ (2008b) Determination of Na and K in biodiesel by flame atomic emission spectrometry and microemulsion sample preparation. J Braz Chem Soc 19:856–861

Chaves ES, de Loos-Vollebregt MTC, Curtius AJ, Vanhaecke F (2011) Determination of trace elements in biodiesel and vegetable oil by inductively coupled plasma optical emission spectrometry following alcohol dilution. Spectrochim Acta Part B At Spectrosc 66:733–739

Chen S, Sun Y, Chao J, Cheng L, Chen Y, Liu J (2016) Dispersive liquid–liquid microextraction of silver nanoparticles in water using ionic liquid 1-octyl-3-methylimidazolium hexafluorophosphate. J Environ Sci 41:211–217

da Silva JDF, da Silva YP, Piatnicki CMS, Böckel WJ, Mendonça CRB (2015) Microemulsions: components, characteristics, potentialities in food chemistry and other applications. Quím Nova 38:1196–1206

Damasceno BPGL, Silva JA, Oliveira EE, Silveira WLL, Araújo IB, Oliveira AG, Egito EST (2011) Microemulsão: um promissor carreador para moléculas insolúveis. Rev Ciênc Farm Básica E Apl 32:9–18

Dancsak SE, Silva SG, Nóbrega JA, Jones BT, Donati GL (2014) Direct determination of sodium, potassium, chromium and vanadium in biodiesel fuel by tungsten coil atomic emission spectrometry. Anal Chim Acta 806:85–90

Dantas TNC, Neto AAD, Moura MCPA, Neto ELB, Forte KR, Leite RHL (2003) Heavy metals extraction by microemulsions. Water Res 37:2709–2717

de Jesus A, Silva MM, Vale MGR (2008) The use of microemulsion for determination of sodium and potassium in biodiesel by flame atomic absorption spectrometry. Talanta 74:1378–1384

de Jesus A, Zmozinski AV, Barbará JA, Vale MGR, Silva MM (2010) Determination of calcium and magnesium in biodiesel by flame atomic absorption spectrometry using microemulsions as sample preparation. Energy Fuels 24:2109–2112

de Oliveira E (2003) Sample preparation for atomic spectroscopy: evolution and future trends. J Braz Chem Soc 14:174–182

de Oliveira AG, Scarpa MV, Correa MA et al (2004) Microemulsões: estrutura e aplicações como sistema de liberação de fármacos. Quim Nova 27:131–138

de Souza RM, Leocádio LG, da Silveira CLP (2008) ICP OES simultaneous determination of Ca, Cu, Fe, Mg, Mn, Na, and P in biodiesel by axial and radial inductively coupled plasma-optical emission spectrometry. Anal Lett 41:1615–1622

DIN EN 14538/2006 (2006) Fat and Oil derivatives. Fatty Acid Methyl Esters (FAME). Determination of Ca, K, Mg and Na content by optical emission spectral analysis with inductively coupled plasma (ICP OES)

dos Santos EJ, Herrmann AB, Chaves ES, Vechiatto WWD, Schoemberger AC, Frescura VLA, Curtius AJ (2007) Simultaneous determination of Ca, P, Mg, K and Na in biodiesel by axial view inductively coupled plasma optical emission spectrometry with internal standardization after multivariate optimization. J Anal At Spectrom 22:1300–1303

Edlund M, Visser H, Heitland P (2002) Analysis of biodiesel by argon–oxygen mixed-gas inductively coupled plasma optical emission spectrometry. J Anal Spectrom 17:232–235

Ezoddin M, Majidi B, Abdi K (2015) Ultrasound-assisted supramolecular dispersive liquid–liquid microextraction based on solidification of floating organic drops for preconcentration of palladium in water and road dust samples. J Mol Liq 209:515–519

Farajzadeh MA, Seyedi SE, Shalamzari MS, Bamorowat M (2009) Dispersive liquid–liquid microextraction using extraction solvent lighter than water. J Sep Sci 32:3191–3200

Farajzadeh MA, Dj Djozan, Bakhtiyar RF (2010) Use of a capillary tube for collecting an extraction solvent lighter than water after dispersive liquid–liquid microextraction and its application in the determination of parabens in different samples by gas chromatography—Flame ionization detection. Talanta 81:1360–1367

Fernández E, Vidal L, Martín-Yerga D et al (2015) Screen-printed electrode based electrochemical detector coupled with ionic liquid dispersive liquid–liquid microextraction and microvolume back-extraction for determination of mercury in water samples. Talanta 135:34–40

Fernández E, Vidal L, Costa-García A, Canals A (2016) Mercury determination in urine samples by gold nanostructured screen-printed carbon electrodes after vortex-assisted ionic liquid dispersive liquid–liquid microextraction. Anal Chim Acta 915:49–55

Fortunato FM, Bechlin MA, Neto JAG, Donati GL, Jones BT (2015) Internal standard addition calibration: determination of calcium and magnesium by atomic absorption spectrometry. Microchem J 122:63–69

Gharehbaghi M, Shemirani F, Baghdadi M (2009) Dispersive liquid–liquid microextraction based on ionic liquid and spectrophotometric determination of mercury in water samples. Int J Environ Anal Chem 89:21–33

Ghisi M, Chaves ES, Quadros DPC, Marques EP, Curtius AJ, Marques ALB (2011) Simple method for the determination of Cu and Fe by electrothermal atomic absorption spectrometry in biodiesel treated with tetramethylammonium hydroxide. Microchem J 98:62–65

Griffin WC (1949) Classification of surface-active agents by "HLB." J Soc Cosmet Chem 311–326

Hardaway C, Sneddon J, Beck JN (2004) Determination of metals in crude oil by atomic spectroscopy. Anal Lett 37:2881–2899

Hashemi B, Zohrabi P, Kim K-H, Shamsipur M, Deep A, Hong J (2017) Recent advances in liquid-phase microextraction techniques for the analysis of environmental pollutants. TrAC Trends Anal Chem 97:83–95

Hoar TP, Schulman JH (1943) Transparent Water-in-Oil Dispersions: the Oleopathic Hydro-Micelle. Nat 152:102–103

Hou W, Xu J (2016) Surfactant-free microemulsions. Curr Opin Colloid Interface Sci 25:67–74

Iqbal J, Carney WA, LaCaze S, Theegala CS (2010) Metals determination in biodiesel (B100) by ICP-OES with microwave assisted acid digestion. Open Anal Chem J 18–26

Jafarvand S, Shemirani F (2011a) Supramolecular–based dispersive liquid–liquid microextraction: a novel sample preparation technique for determination of inorganic species. Microchim Acta 173:353–359

Jafarvand S, Shemirani F (2011b) Supramolecular-based dispersive liquid–liquid microextraction: a novel sample preparation technique utilizes coacervates and reverse micelles. J Sep Sci 34:455–461

Jafarvand S, Shemirani F (2011c) Supramolecular-based dispersive liquid–liquid microextraction: determination of cadmium in water and vegetable samples. Anal Methods 3:1552–1559

Jain S, Sharma MP (2011) Correlation development for effect of metal contaminants on the oxidation stability of Jatropha curcas biodiesel. Fuel 90:2045–2050

Khodakiya AS, Chavada JR, Jivani NP, Patel BN, Khodakiya MS, Ramoliya AP (2012) Microemulsions as enhanced drug delivery carrier: an overview. Am J Pharmtech Res 206–226

Klossek ML, Touraud D, Zemb T, Kunz W (2012) Structure and solubility in surfactant-free microemulsions. ChemPhysChem 13:4116–4119

Knothe G (2006) Analyzing biodiesel: standards and other methods. J Am Oil Chem Soc 83: 823–833

Knothe G, Dunn RO (2003) Dependence of oil stability index of fatty compounds on their structure and concentration and presence of metals. J Am Oil Chem Soc 80:1021–1026

Korn M, Santos D, Welz B, Vale M, Teixeira A, Lima D, Ferreira S (2007) Atomic spectrometric methods for the determination of metals and metalloids in automotive fuels—A review. Talanta 73:1–11

Lisboa MT, Clasen CD, Vellar DC de S, Oreste EQ, Saint'Pierre TD, Ribeiro AS, Vieira MA (2013) An easy and fast procedure for the determination of Ca, Cu, Fe, Mn, Mg, Na, K and Si in biodiesel by ICP OES using emulsification as sample preparation strategy. J Braz Chem Soc 143–151

Liu Y, He M, Chen B, Hu B (2015) Simultaneous speciation of inorganic arsenic, selenium and tellurium in environmental water samples by dispersive liquid liquid microextraction combined with electrothermal vaporization inductively coupled plasma mass spectrometry. Talanta 142:213–220

Liu Y, He M, Chen B, Hu B (2016) Ultra-trace determination of gold nanoparticles in environmental water by surfactant assisted dispersive liquid liquid microextraction coupled with electrothermal vaporization-inductively coupled plasma—mass spectrometry. Spectrochim Acta Part B At Spectrosc 122:94–102

Llorent-Martínez EJ, Ortega-Barrales P, Fernández-de Córdova ML, Domínguez-Vidal A, Ruiz-Medina A (2011) Investigation by ICP-MS of trace element levels in vegetable edible oils produced in Spain. Food Chem 127:1257–1262

Lobo FA, Goveia D, de Oliveira AP, Pereira-Filho ER, Fraceto LF, Filho NLD, Rosa AH (2009) Comparison of the univariate and multivariate methods in the optimization of experimental conditions for determining Cu, Pb, Ni and Cd in biodiesel by GFAAS. Fuel 88:1907–1914

Lobo FA, Goveia D, Oliveira AP, Romão LPC, Fraceto LF, Dias Filho NL, Rosa AH (2011) Development of a method to determine Ni and Cd in biodiesel by graphite furnace atomic absorption spectrometry. Fuel 90:142–146

López-García I, Vicente-Martínez Y, Hernández-Córdoba M (2014) Determination of cadmium and lead in edible oils by electrothermal atomic absorption spectrometry after reverse dispersive liquid–liquid microextraction. Talanta 124:106–110

López-García I, Briceno M, Vicente-Martínez Y, Hernández-Córdoba M (2015) Rapid screening of water soluble arsenic species in edible oils using dispersive liquid–liquid microextraction. Food Chem 167:396–401

López-Jimenez FJ, Rubio S, Pérez-Bendito D (2008) Single-drop coacervative microextraction of organic compounds prior to liquid chromatography theoretical and practical considerations. J Chromatogr A 1195:25–33

Lyra FH, Carneiro MTWD, Brandão GP, Pessoa HM, de Castro EV (2010) Determination of Na, K, Ca and Mg in biodiesel samples by flame atomic absorption spectrometry (F AAS) using microemulsion as sample preparation. Microchem J 96:180–185

Mahmoud C, Mohadese HPZ, Hossein AZM (2014) Dispersive Liquid-Liquid microextraction based on solidification of floating organic drop combined with flame atomic absorption spectrometry for preconcentration and determination of Thallium(III) in water samples. Iran J Chem Chem Eng 33:59–66

Mandlate JS, Soares BM, Seeger TS, Dalla Vechia P, Mello PA, Flores EMM, Duarte FA (2017) Determination of cadmium and lead at sub-ppt level in soft drinks: an efficient combination between dispersive liquid-liquid microextraction and graphite furnace atomic absorption spectrometry. Food Chem 221:907–912

Martinis EM, Wuilloud RG (2016) Enhanced spectrophotometric detection of Hg in water samples by surface plasmon resonance of Au nanoparticles after preconcentration with vortex-assisted liquid-liquid microextraction. Spectrochim Acta A Mol Biomol Spectrosc 167:111–115

Maya F, Horstkotte B, Estela JM, Cerdá V (2014) Automated in-syringe dispersive liquid-liquid microextraction. Trends Anal Chem 59:1–8

Merck KGaA, Darmstadt, Germany. http://www.sigmaaldrich.com/materials-science/material-science-products.html?TablePage=22686648, Accessed 01 Nov 2017

Mitchell DJ, Ninham BW (1981) Micelles, vesicles and microemulsions. J Chem Soc, Faraday Trans 277:601–629

Mohammadi SZ, Baghelani YM, Mansori F, Shamspur T, Afzali D (2012) Dispersive liquid-liquid microextraction for the simultaneous separation of trace amounts of zinc and cadmium ions in water samples prior to flame atomic absorption spectrometry determination. Quím Nova 35:198–202

Molaakbari E, Mostafavi A, Afzali D (2011) Ionic liquid ultrasound assisted dispersive liquid–liquid microextraction method for preconcentration of trace amounts of rhodium prior to flame atomic absorption spectrometry determination. J Hazard Mater 185:647–652

Moral A, Sicilia MD, Rubio S (2009) Determination of benzimidazolic fungicides in fruits and vegetables by supramolecular solvent-based microextraction/liquid chromatography/ fluorescence detection. Anal Chim Acta 650:207–213

Ni P, Hou W-G (2008) A Novel Surfactant-free Microemulsion System: *N*, *N*-Dimethyl Formamide/Furaldehyde/$H_2O$. Chin J Chem 26:1335–1338

Nunes LS, Barbosa JTP, Fernandes AP et al (2011) Multi-element determination of Cu, Fe, Ni and Zn content in vegetable oils samples by high-resolution continuum source atomic absorption spectrometry and microemulsion sample preparation. Food Chem 127:780–783

Ojeda CB, Rojas FS, Pavón JMC (2012) Determination of Cobalt in Food, Environmental and Water Samples with Preconcentration by Dispersive Liquid-Liquid Microextraction. Am J Anal Chem 3:125–130

Ozdemir C, Saçmaci S, Kartal S, saçmaci M (2014) Determination of gold and palladium in environmental samples by FAAS after dispersive liquid–liquid microextraction pretreatment. J Ind Eng Chem 20:4059–4065

Pelizzette E, Pramauro E (1985) Analytical Applications of organized molecular assemblies. Anal Chim Acta 169:1–29

Pereira FM, Zimpeck RC, Brum DM, Cassella RJ (2013) Novel extraction induced by emulsion breaking as a tool for the determination of trace concentrations of Cu, Mn and Ni in biodiesel by electrothermal atomic absorption spectrometry. Talanta 117:32–38

Pereira FM, Brum DM, Lepri FG, Cassella RJ (2014) Extraction induced by emulsion breaking as a tool for Ca and Mg determination in biodiesel by fast sequential flame atomic absorption spectrometry (FS-FAAS) using Co as internal standard. Microchem J 117:172–177

Pillay AE, Elkadi M, Fok SC et al (2012) A comparison of trace metal profiles of neem biodiesel and commercial biofuels using high performance ICP-MS. Fuel 97:385–389

Pouyan M, Bagherian G, Goudarzi N (2016) Determination of ultra-trace palladium (II) in water, soil, and food samples by dispersive liquid-liquid microextraction-atomic absorption spectrometry using 2-mercaptobenzimidazole as a complexing agent. Microchem J 127:46–51

Quadros DPC, Rau M, Idrees M, Chaves ES, Curtius AJ, Borges DLG (2011) A simple and fast procedure for the determination of Al, Cu, Fe and Mn in biodiesel using high-resolution continuum source electrothermal atomic absorption spectrometry. Spectrochim Acta Part B At Spectrosc 66:373–377

Raposo JD, Costa LM, Barbeira PJS (2014) Simultaneous determination of Na, K and Ca in biodiesel by flame atomic emission spectrometry. J Braz Chem Soc 26:147–155

Rezaee M, Assadi Y, Milani Hosseini M-R et al (2006) Determination of organic compounds in water using dispersive liquid–liquid microextraction. J Chromatogr A 1116:1–9

Rezaee M, Yamini Y, Faraji M (2010) Evolution of dispersive liquid-liquid microextraction method. J Chromatogr A 1217:2342–2357

Rosa FC, Duarte FA, Paniz JNG et al (2015) Dispersive liquid–liquid microextraction: an efficient approach for the extraction of Cd and Pb from honey and determination by flame atomic absorption spectrometry. Microchem J 123:211–217

Saint'Pierre TD, Dias LF, Maia SM, Curtius AJ (2004) Determination of Cd, Cu, Fe, Pb and Tl in gasoline as emulsion by electrothermal vaporization inductively coupled plasma mass spectrometry with analyte addition and isotope dilution calibration techniques. Spectrochim Acta Part B at Spectrosc 59:551–558

Sakai T (2008) Surfactant-free emulsions. Curr Opin Colloid Interface Sci 13:228–235

Sarin A, Arora R, Singh NP, Sharma M, Malhotra RK (2009) Influence of metal contaminants on oxidation stability of Jatropha biodiesel. Energy 34:1271–1275

Sarin A, Arora R, Singh NP, Sarin R, Malhotra RK (2010) Oxidation stability of palm methyl ester: effect of metal contaminants and antioxidants. Energy Fuels 24:2652–2656

Schulman JH, Stoeckenius W, Prince LM (1959) Mechanism of formation and structure of micro emulsions by electron microscopy. J Phys Chem 63:1677–1680

Shamsipur M, Fattahi N, Assadi Y, Sadeghi M, Sharafi K (2014) Speciation of As(III) and As(V) in water samples by graphite furnace atomic absorption spectrometry after solid phase extraction combined with dispersive liquid–liquid microextraction based on the solidification of floating organic drop. Talanta 130:26–32

Shirkhanloo H, Ghazaghi M, Eskandari MM (2016) Cloud point assisted dispersive ionic liquid - liquid microextraction for chromium speciation in human blood samples based on isopropyl 2-[(isopropoxycarbothiolyl)disulfanyl] ethane thioate. Anal Chem Res 10:18–27

Shrivas K, Jaiswal NK (2013) Dispersive liquid–liquid microextraction for the determination of copper in cereals and vegetable food samples using flame atomic absorption spectrometry. Food Chem 141:2263–2268

Silva JSA, Chaves ES, Santos ÉJ dos, Saint'Pierre TD, Frescura VLA, Curtius AJ (2010) Calibration techniques and modifiers for the determination of Cd, Pb and Tl in biodiesel as microemulsion by graphite furnace atomic absorption spectrometry. J Braz Chem Soc 21: 620–626

Smith GD, Donelan CE, Barden RE (1977) Oil-continuous microemulsions composed of hexane, water, and 2-propanol. J Colloid Interface Sci 60:488–496

Song F, Xu J, Hou WG (2010) Surfactant-free oil/water and bicontinuous microemulsion composed of benzene, ethanol and water. Chin Chem Lett 21:880–883

Thompson JE, Davidow LW (2013) A Prática Farmacêutica na Manipulação de Medicamentos, 3rd edn. Artmed, Porto Alegre

Viana C, Bohrer D, de Carvalho LM, do Nascimento PC, da Rosa MB (2014) Emulsified systems for metal determination by spectrometric methods. TrAC Trends Anal Chem 53:49–59

Vicentino PO, Cassella RJ (2017) Novel extraction induced by microemulsion breaking: a model study for Hg extraction from Brazilian gasoline. Talanta 162:249–255

Vieira MA, de Oliveira LCC, Gonçalves RA, de Souza V, de Campos RC (2009) Determination of as in vegetable oil and biodiesel by graphite furnace atomic absorption spectrometry. Energy Fuels 23:5942–5946

Woods GD, Fryer FI (2007) Direct elemental analysis of biodiesel by inductively coupled plasma–mass spectrometry. Anal Bioanal Chem 389:753–761

Xiao-Huan Z, Qiu-Hua W, Mei-Yue Z, Guo-Hong X, Zhi W (2009) Developments of dispersive liquid-liquid microextraction technique. Chin J Anal Chem 37(2):161–168

Xu J, Yin A, Zhao J, Li D, Hou W (2013) Surfactant-free microemulsion composed of oleic acid, *n* -Propanol, and $H_2O$. J Phys Chem B 117:450–456

Yaakob Z, Narayanan BN, Padikkaparambil S, Unni KS, Akbar PM (2014) A review on the oxidation stability of biodiesel. Renew Sustain Energy Rev 35:136–153

Yan H, Wang H (2013) Recent development and applications of dispersive liquid–liquid microextraction. J Chromatogr A 1295:1–15

Yousefi SR, Ahmadi SJ (2011) Development a robust ionic liquid-based dispersive liquid-liquid microextraction against high concentration of salt combined with flame atomic absorption spectrometry using microsample introduction system for preconcentration and determination of cobalt in water and saline samples. Microchim Acta 172:75–82

Zgoła-Grześkowiak A, Grześkowiak T (2011) Dispersive liquid-liquid microextraction. TrAC Trends Anal Chem 30:1382–1399

# Chapter 7
# Alternative Uses for Biodiesel Byproduct: Glycerol as Source of Energy and High Valuable Chemicals

**Cauê Alves Martins, Pablo Sebastián Fernández and Giuseppe Abíola Camara**

**Abstract** Glycerol was firstly faced as a residue, since it is massively produced from the transesterification of vegetable oils and animal fat, corresponding to roughly 10% of the total amount of biodiesel. The recent high availability of glycerol has decreased its price and increased the risks of environmentally inadequate disposal. Now, this small chain alcohol is faced as a powerful alternative for energy conversion and production of chemicals with commercial interest. Its three hydrated carbons make this organic a noticeable substrate to produce other carbonyl compounds and to be used to produce energy in electrochemical devices collectively known as direct alcohol fuel cells. The development in the electrocatalysis field opened up new strategies to convert glycerol into power and chemicals, by using fuel cells and electrolyzer reactors. In both systems, the efficiency of the process depends on several aspects, as the medium, applied potential, and mainly on the surface reaction taking place at the interface between solution and electrode. In this chapter, we discuss the use of glycerol in these electrochemical systems and illustrate some experimental results regarding fundamental and applied science. Namely, we describe some advances in the understanding of the glycerol electro-oxidation reaction interpreted by spectroscopy and chromatographic techniques. Novel nanomaterials currently applied to improve the catalysis of the reaction are also shown. Moreover, we comment some results regarding fuel cells, microfluidic fuel cells, and electrolyzers fed by glycerol and the perspectives and challenges of its use.

---

C. A. Martins (✉)
Faculty of Exact Sciences and Technology, Federal University of Grande Dourados, 79804-970 Dourados, MS, Brazil
e-mail: cauealvesmartins@gmail.com; cauemartins@ufgd.edu.br

P. S. Fernández
Institute of Chemistry, Universidade Estadual de Campinas. Cidade Universitária "Zeferino Vaz", Barão Geraldo, CEP 13083-970 Campinas, SP, Brazil

G. A. Camara
Institute of Chemistry, Federal University of Mato Grosso do Sul, C.P. 549, 79070-900 Campo Grande, MS, Brazil

 
M. Trindade (ed.), *Increased Biodiesel Efficiency*, Green Energy and Technology,
https://doi.org/10.1007/978-3-319-73552-8_7

## 7.1 Introduction

The previous chapters of this book made clear the importance of the use of biodiesel as a source for economic and environmental friendly energy production. The transesterification of animal fats or vegetal oils, named alcoholysis, is a wide known method to produce biodiesel, forming esters and glycerol. Such reaction yields roughly 10% of glycerol (w/w). In this sense, the increasing in biodiesel production is resulting in a surplus of glycerol in the market, which decreases the price of the commodity and raises the chances of inadequate disposal.

Tropical countries with adequate climate for oleaginous growth, like Brazil, are continuously producing biodiesel in industrial scale, which is being used for stationary engines and vehicles (Oliveira et al. 2016). Part of the regular diesel is replaced by biodiesel. In Brazil, a presidential decree increased to 8% the amount of biodiesel in diesel composition in 2017 and will reach 10% until 2019. In 2015, the country produced 390 million liters of glycerol, which is far beyond its annual demand of use. Such biodiesel production was bigger than the German's and Argentine's, just behind the USA (Oliveira et al. 2016). Brazilian government expects to produce 7 billion litters of biodiesel in 2020, which corresponds to 700 million liters of glycerol in the market. Although there is demand for glycerol in several industry sectors, the availability is so high that might compromise the environmentally clean characteristic of using biodiesel, because of the lack of concern regarding the storage and discarding of glycerol. In this sense, the way to keep inserting the biodiesel fuel inside the economic matrix of energy production is to find ways of using glycerol, which must be economically viable and environmental friendly.

In the context of energy conversion, the advances in science and technologies of fuel cells opened up the possibility of using alcohols as fuels to produce energy efficiently with low environmental impact (Oliveira et al. 2016). These devices, also named direct alcohol fuel cells (DAFCs), convert the electrical current produced through two half-electrochemical reactions into available work. DAFCs are basically composed of anode, cathode, and an ionic exchange layer, usually a polymeric membrane. The alcohol must be electro-oxidized at the anode while an oxidant, usually oxygen from air, is electroreduced at the cathode, generating electrical current. The power output is proportional to the equivalent amount of electrons extracted from the fuel; therefore, the power output is maximized when the reaction is complete. Glycerol contains three hydrated carbons in its chain, which allied with its availability, increases its potential use as fuel for DAFC. However, there are challenges to be faced before making this technology commercially available.

The anodes of fuel cells are composed of metallic nanoparticles (NPs), commercially available as Pt NPs dispersed on carbon supports (Pt/C). Although Pt is active for fuel cells fed by hydrogen, it does not reach the required activity for alcohol electro-oxidation due to the limited ability for C–C cleavage in low potentials, which prevents total conversion toward $CO_2$, decreasing the overall efficiency of the device (Oliveira et al. 2016; Martins et al. 2014b). Moreover, the anode made of Pt is the most expensive part of a fuel cell. Therefore, much effort has been spent toward the synthesis of new nanomaterials with enhanced activity and with lower content of Pt, usually by replacing it partially or totally for other d-group metals (Figueiredo et al. 2015; Zanata et al. 2016; Fontes et al. 2016; Mello et al. 2017). In addition to the synthesis of new NPs, research groups have been dedicating to the understanding of the interface reaction, performing glycerol electro-oxidation reaction (GEOR) using well-ordered surfaces (Fernández et al. 2015a, b, 2016; Garcia et al. 2016), to shed a light on the influence of the superficial atomic arrangement of the catalyst on the electrochemical performance.

Besides its potential use for energy conversion, glycerol has been considered a matrix reactant for electrosynthesis, when glycerol is converted to other chemicals by using electrochemical systems, named electrolysers (Roquet et al. 1994a; Kishida et al. 2005; Yuksel et al. 2010). Similar to fuel cell, electrolyzers are composed of nanomaterials used as catalysts and the reaction takes place in an electrode/solution interface. This approach is based on the fact that there are organic compounds with high market prices, which might be synthesized from glycerol. The selectivity of the reaction is based on multiple parameters, among which the catalyst is the most important. The basic difference between fuel cells fed by glycerol and chemical conversion of glycerol to high valuable compounds is the thermodynamics involved. Fuel cells are thermodynamically opened systems considered galvanic cells which continuously convert supplied fuel and oxidant into electrical energy, while electrolyzers convert glycerol into other chemicals as a result of an applied potential. In this sense, the investigations of glycerol electro-oxidation reactions taking place at an electrode/solution interface give great contribution to the development of both devices.

The overall goal of this chapter is to provide a general overview of the glycerol electro-oxidation reaction in bulk and nanoparticle electrodes in support of future researches, materials development, products development, and to guide the researcher to face some challenges for the use of glycerol in energy conversion and electrosynthesis. This chapter should not be read as a review because it is limited to only a fraction of the works regarding the use of glycerol in electrochemical systems. The scope is limited to some reports addressing glycerol electro-oxidation in half-cell measurements, experiments using spectroelectrochemical and chromatographic measurements, and some examples of applied systems. This chapter is outlined as follows: (1) an illustrative overview of fuel cell and electrolyzer fed by glycerol; (2) proposed pathways for the reaction and the contributions of bulk and nanoparticle electrodes, including examples of some breakthrough contributions in the field; (3) examples of applied systems, which intend to shed a light on ongoing and emerging real devices; and (4) an outlook for future entice students, researchers, and industrials.

## 7.1.1 *Fuel Cells*

To the moment, there are no fuel cells fed by glycerol running in vehicles or in mobile devices, and the few devices in operation are still prototypes, but owing to some attractive features (discussed latter in this chapter), many studies involving the electro-oxidation of glycerol in fuel cells and half cells are emerging from the scientific community in the last years (Bianchini and Shen 2009; Simões et al. 2011, 2012; Zhang et al. 2012a; Falase et al. 2012; Fernández et al. 2012a; Marchionni et al. 2013; Holade et al. 2013; Martins et al. 2014b; Bott-Neto et al. 2014; Zanata et al. 2016; Huang et al. 2016).

A fuel cell is an electrochemical device that spontaneously generates electrical work from two chemical reactions that are interconnected. As these reactions depend on each other, they are called half-reactions and this interdependence means that the same electrons that are donated from the first half-reaction (an oxidation) travel the cell throughout an electrical conductor and enter the second compartment, where they reduce reactive species in the second half-reaction (a reduction).

Figure 7.1 shows how a fuel cell fed by glycerol operates in a generic way. A solution containing glycerol enters the first compartment (the anode of the cell), where the alcohol is oxidized in a proper catalyst (a substance that promotes the reaction, major details will be given later). The exact number of electrons released by each molecule depends on the products formed during the oxidation step (Zhang et al. 2012a; Simões et al. 2012; Marchionni et al. 2013), but for now, it is enough to know that once the molecules suffer oxidation, they release protons ($H^+$) and

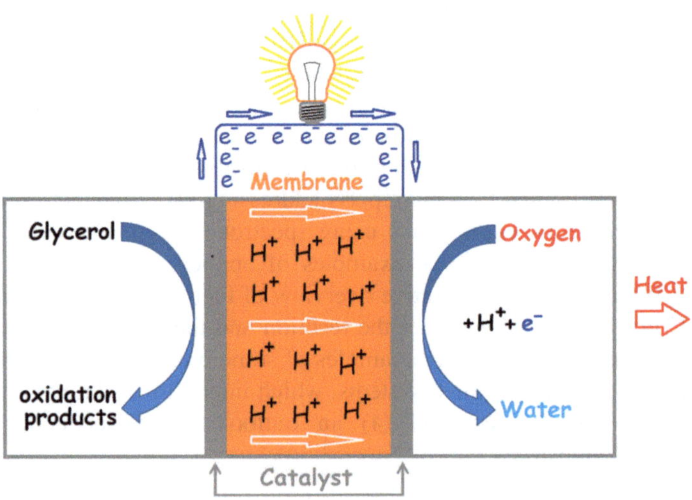


**Fig. 7.1** Operating principle of a fuel cell fed by glycerol/oxygen. As glycerol oxidizes in the anode (left side), protons ($H^+$), and electrons are released. Protons cross a proton-exchange membrane while electrons travel outside the cell and perform electrical work. At the end, protons and electrons are combined in the cathode (right side) to generate water

electrons. Protons cross the cell via a polymeric membrane specifically designed for this purpose (Wang et al. 2011). At the same time, the electrons released by the alcohol exit the cell through the so-called external circuit and in their way toward the other compartment they can perform electrical work, after what they enter in the cell by the opposite electrode (the cathode) and are combined to those protons that have crossed the membrane and oxygen (captured from air, for instance) to generate water (Wang et al. 2011; Babir et al. 2012). More recently, there are also fuel cells that operate with alkaline membranes, but the operation principle is basically the same, except that hydroxyl species cross the membrane in the opposite direction. Further details about the half-reactions involved will be given later in this chapter.

Interestingly, both half-reactions can happen at the same place if the molecules involved in the process (details about what we already know from the mechanism will be discussed later) are left in the same vessel and the conditions are such that the reaction can take place. However, in this case, the electrons are directly transferred from one molecule to another, which means that they cannot be used to generate electricity. In other words, a fuel cell is a device that splits a chemical reaction into two parts that are spatially separated (the half-reactions of oxidation and reduction) and, through electronic conduction, takes advantage of the journey of the electrons to cause them to generate work. As in any real process, some energy is lost as heat, but the fuel cell can be designed so that even part of this heat can be reused as work, which in principle allows theoretical yields exceeding 83% to be obtained (Pilatowsky et al. 2011).

When compared to classical fuel cells (those fed by hydrogen and oxygen), direct glycerol fuel cells present some advantages. First, glycerol is a low volatile, non-toxic, and non-explosive liquid (Gupta and Kumar 2012), which implies that the storage conditions are simpler than those required for hydrogen. Another important factor is the price of the two fuels. Owing to the dramatic increase in the production of biodiesel worldwide, the excess of glycerol on the global market has made it increasingly inexpensive (current values are around 500 $/ton) (Bagnato et al. 2017). On the other hand, the costs for production of hydrogen from water electrolysis were estimated as $\sim$3.3 $/kg (Independent Review Panel Summary Report 2009) (there are methods of production of hydrogen less expensive than electrolysis; they include photoelectrochemical water splitting, steam reforming, fermentation, etc., but the presence of impurities seriously compromises the energetic yield of the fuel cell and makes its use impracticable). These values are 6.6 times higher than those observed for glycerol. When both fuels are normalized by their respective theoretical energy densities, the costs of energy tie at 0.1 US$ per KWh. Considering the costs of compressing (in case of hydrogen), storage, and dispensing of the fuels (Independent Review Panel Summary Report 2009), the use of glycerol in fuel cells presents a clear advantage over hydrogen in economic terms.

At this point, some considerations regarding the technology of PEM fuel cells can be summarized. Concerning hydrogen fuel cells, the kinetics associated with the corresponding hydrogen electro-oxidation is fast, particularly in acidic environment (Hamann et al. 2007). The problems here are chiefly related to the costs of the technology—as fuel cells use mostly platinum as catalyst (Song 2002)—and the costs of adapting an worldwide infrastructure to store and distribute hydrogen, which is not an straightforward task, since hydrogen is a tiny molecule, and therefore very diffusive, besides the fact that it is very explosive. That is, we must take into account the costs of the technology itself and add to it the costs of safety. However, despite these limitations, we know how to oxidize hydrogen efficiently. On the other hand, when we talk about direct glycerol fuel cells, we are facing a different kind of trouble: Despite the fact that the storage and distribution of alcohols is much easier to build, since they are liquids (for instance, in Brazil, these facilities already exist for ethanol and could be adapted for glycerol), the truth is we still are not able to oxidize glycerol efficiently. The reasons for such low efficiency are closely related to reactions that are generally branched and sluggish (Martins et al. 2014b). The state of the art of the mechanisms involved will be discussed later.

### *7.1.2 Electrolyzers*

Another way to use glycerol in electrochemical systems is to reverse the principle of a fuel cell—that spontaneously generates electrical power while promotes the occurrence of an electrochemical reaction—by making use of an electrolyzer. In this kind of device, energy is consumed to promote the electrolysis of the alcohol by a couple of half-reactions whose global reaction is not spontaneous. Specifically during the electrolysis of glycerol, the global theoretical reaction of decomposition is given by Eq. 7.1 (Simões et al. 2012; González-Cobos et al. 2016):

$$CH_2OH{-}CHOH{-}CH_2OH + 3\,H_2O \rightarrow 3\,CO_2 + 7\,H_2 \tag{7.1}$$

This reaction means that glycerol can be theoretically used to obtain hydrogen. Additionally, the gases represented in Eq. 7.1 are produced in separated compartments in the electrolyzer (see Fig. 7.2 for details), which means that hydrogen (and $CO_2$) with high levels of purity can be obtained, thus circumventing the problems caused by the impurities typical of hydrogen obtained from fossil resources (González-Cobos et al. 2016). Also, for ethanol in alkaline environment, it has been demonstrated that hydrogen can be obtained by consuming just about one-third of the energy required by the electrolysis of water (Bambagioni et al. 2010), which minimizes the cost problems already mentioned in the previous section.

Figure 7.2 shows the working principle of a proton exchange membrane electrolysis cell. The reader is invited to compare this figure with Fig. 7.1 in the previous section to see how similar both systems are. Indeed, the reactions that take place at the anode (the side fed by glycerol) are of the same nature. The other electrode (the cathode) is the one who determines if the oxidation is spontaneous or

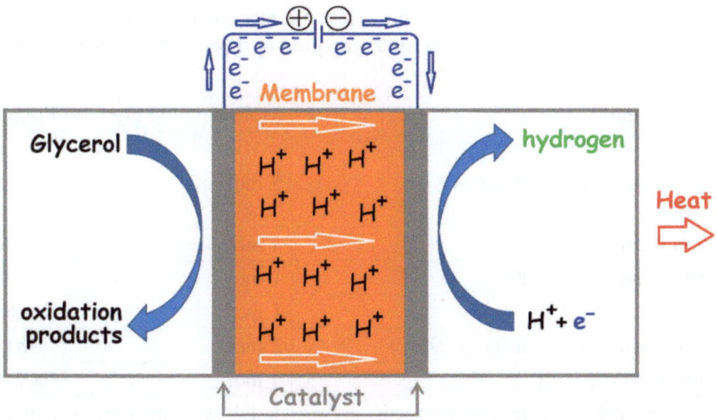


**Fig. 7.2** Operating principle of an electrolyzer fed by glycerol. Glycerol oxidizes in the anode (left side) owing to the imposition of an external voltage, while protons ($H^+$) and electrons are released. Protons cross a proton-exchange membrane while electrons are forced to travel toward the cathode. Protons and electrons are combined in the cathode (right side) to generate gaseous hydrogen

not, once the occurrence of a reaction in nature is given by a thermodynamic parameter called standard cell potential (formally described as $E^o$), that in turns depends on the difference of voltage between cathode and anode. In our comparison, the replacement of (a) a cathode that reduces oxygen by (b) a second one that reduces protons implies that the oxidation of glycerol is spontaneous in (a) and generates electric power and non-spontaneous, power-driven in (b). Further details about the thermodynamics of both processes are beyond the scope of this chapter, but the reader is invited to consult references (Simões et al. 2012), (González-Cobos et al. 2016) and (Bambagioni et al. 2010) to deepen the knowledge on this subject.

Besides clean $H_2(g)$ at the cathode, the configuration showed in Fig. 7.2 allows the production (synthesis) of carbonyl compounds. Since glycerol might undergo toward a myriad of carbonyl compounds, an electrolyzer can be used to massively synthesize glycerol-derived products. Different from classic wet chemical synthesis, catalysts, experimental condition and applied potential can be adjusted to tune the products (Horn et al. 2016). Most of these substances are high-added-value chemical compounds. Table 7.1 shows some of the reaction products obtained during the electrolysis of glycerol and their respective prices purchased from Sigma-Aldrich. The reactions and experimental conditions for production of each substance are described in reference (Coutanceau and Baranton 2016). The price of glycerol is also included for ends of comparison, as well as the number of moles of hydrogen produced per mole of glycerol consumed, as described in (Coutanceau and Baranton 2016).

Table 7.1 shows that the cogeneration of chemicals during the production of gaseous hydrogen in glycerol electrolyzers can be an extremely attractive strategy

**Table 7.1** Prices of some of the reaction products of intact chain obtained during electrolysis of glycerol, with the corresponding moles of hydrogen produced per mole of glycerol consumed in an electrolyser

| Chemical compound | Mole $H_2$/mole glyc. | Price (US$ $g^{-1}$) |
|---|---|---|
| Glycerol | – | 0.33 |
| [a]Sodium mesoxalate monohydrate (purity $\geq$ 98%) | 5 mol | 5.95 |
| Tartronic acid (purity $\geq$ 97%) | 4 mol | 6.53 |
| Glyceraldehyde (purity $\geq$ 90%) | 1 mol | 86.50 |
| Dyhidroxyacetone | 1 mol | 1389.60 |
| [a]Glyceric acid sodium salt (purity $\geq$ 95%) | 2 mol | 3812.00 |
| Hydroxypyruvic acid (purity $\geq$ 95%) | 4 mol | 18,740.00 |

[a]In these cases, the prices of the cheapest salts are presented, once the corresponding acids were not found in the online catalogue

from the commercial point of view, since some of these substances present a market price several orders of magnitude higher than glycerol, which certainly compensates the energy expended during the electrolysis. Moreover, a hydrogen/oxygen fuel cell can be theoretically used in conjunction with an electrolyzer fed by glycerol to produce energy with no need of storing hydrogen, which means more safety and reduced costs with gas pressurization in these systems.

Despite the positive aspects discussed in the last paragraph, one of the main challenges in alcohol electrolyzers is the need to drive the oxidation reaction to specific routes (and products). Otherwise, a mixture of compounds will be obtained, whose separation processes of which may be very costly, and eventually override the advantages discussed herein. The ways of doing so have been intensely debated in the specialized literature during the last few years and will be commented below.

The arguments raised here make clear that glycerol electro-oxidation provides a unique opportunity to allow the production of energy and high valuable compounds in parallel. Considering the investigation of electrolysis of glycerol in a non-spontaneously system, where some applied potential is used to convert glycerol into carbonyl products, similar products might be obtained whether the fuel cell is operating at similar condition (i.e., same solution and catalysts) and equivalent open circuit potential to the voltage applied in an electrolyzer.

In principle, a major inconvenient in the electrolysis as well as in feeding fuel cells with molecules relatively complex as alcohols—glycerol has a three-carbon chain—is the complexity of the electro-oxidation mechanism. The exact nature of the products formed during the oxidation steps depends on the way how the molecule and the surface interact. In this sense, parameters as chemical composition of the surface and even aspects in atomic scale (for instance, how a particular molecule is bind to a surface) can determine the chemical nature of the products. The output is miscellanea of parallel reactions that produce several substances. From an energetic viewpoint, each substance will release a different number of electrons and protons, which means that the chemical nature of the product will determine the efficiency of the global process.

### 7.1.3 The Catalyst

All the reactions involved in fuel cells and in electrolizers occur over a surface (for example, metallic NPs). The surface is called "catalyst," and the chemical reaction occurring on the catalyst is classified as a "catalyzed reaction." The catalyst is probably the most important component in every catalyzed process as it is able to drive the reaction through different pathways by changing the velocity of the parallel and consecutive reactions that usually take place during the transformation of any reactant. By just changing the catalyst, we can go from a fuel cell not delivering energy to one capable of moving a bus or from an electrolizer producing a big mixture of products (non-selective reaction) to one able to generate only one pure substance.

The processes mentioned before occur as it is shown schematically in Fig. 7.3. This figure shows, in a simplified way, a glycerol molecule reacting on three different catalysts (for example, Au NPs, Pt NPs, and Ni NPs). Lets imagine the reactant interacts with the surface (as for example the catalyst in blue) to form three different products (A, B, and C). When we change the catalyst from the blue to the yellow one, glycerol interacts with the catalyst in a different manner. In some way (that can be very difficult to find out), the reaction forming the product A is speeded up, whereas the production of C does not change and the product B is not observed, which means that the reaction pathway that renders this product is inhibited. In the last example, again the three products are simultaneously generated, but the change of the catalyst has clearly increased the formation of B. Imagine we are thinking in an industrial process to transform glycerol in product B and we have to choose one of our materials. In this case, if we consider only the points discussed before (ignoring the price, for instance), the green catalyst is the most interesting, once it is

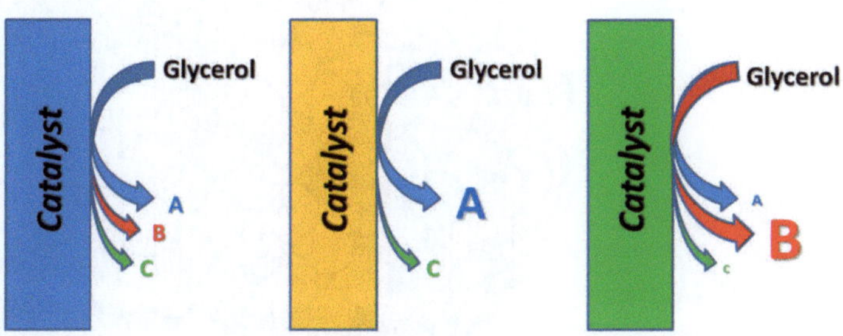


**Fig. 7.3** Illustrative scheme of the influence of the catalyst on the kinetics and selectivity of glycerol-derived products. Bigger arrows represent faster kinetics. The bigger the letter is the higher the production of the molecule

the one providing a faster reaction and a selective production of B molecules compared to A and C, which facilitates the purification process.

Considering our previous discussion, we can define a catalyst as a surface that interacts with reactants, reaction intermediates, and products, changing simultaneously the kinetics by decreasing the overall activation energy of the reactions and tuning the selectivity of a process. In this context, the ability of a researcher or industrial to produce nanomaterials with very high area/mass ratio is strongly desired for catalytic processes, due to the fact that for a given material, the rate of the conversion of a reactant is proportional to the area of the catalyst. Today, metals, oxides, alloys, etc., can be easily produced in nanometric size. There are a myriad of methods to produce Au, Ag, Pt, etc. NPs with high specific surface areas (Solla-Gullón et al. 2013; Sau and Rogach 2010), which have generated an explosion in the amount of scientific data.

Figure 7.4 shows a model of a NP containing atoms labeled with different colors. If we look at any atom in the green region, we will see that it is in contact with other six atoms in the same plane. This configuration is known as (111). The atoms in the blue regions have four close neighbors, having a configuration known as (100), and those between these regions have even less atoms in contact with them, reminding a (110) configuration. Apart from this structural point of view, it is important to know that the chemical properties of the atoms strongly depend of the number on neighbors (called coordination number). Thus, even if all the atoms are Pt, for instance, their physicochemical properties will vary depending on the region of the nanoparticle where they are. Therefore, a given molecule, for example glycerol, will interact in a different way in different regions of the nanoparticle, and the electro-oxidation of the molecule will have different selectivity/rate depending on which part of the nanoparticle the molecule reached.

**Fig. 7.4** Illustrative model of a nanoparticle

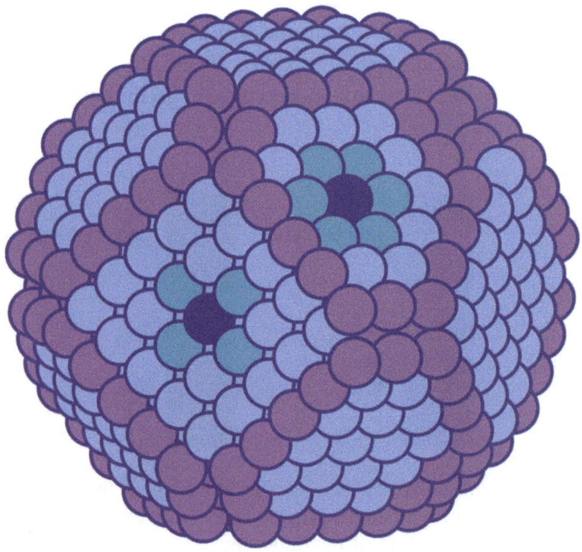

It is important to know that the complexity mentioned before is far beyond from the complexity of real systems where we have generally millions of NPs, with a distribution of different sizes and shapes. These facts multiplied the configurations described in Fig. 7.4. Moreover, Pt NPs do not show relevant results in terms of current output and selectivity (as shown in the next sections), so it is necessary to modify the catalyst by synthesizing alloys and decorated multi-metallic NPs for glycerol electro-oxidation. These complex NPs are dispersed on carbon support, which can be amorphous, nanotubes, or even graphene-based materials (Fernández et al. 2012a; Martins et al. 2014a). The specific discussion regarding each of these topics is beyond the scope of this chapter. Here, we show a general view of the glycerol electro-oxidation reaction and the influence of the metallic content of the catalyst on the reaction.

## 7.2 Glycerol Electro-Oxidation Reaction

Among the researches involving glycerol, those regarding the investigation of the surface reaction taking place at the electrode/solution interface greatly enhance the understanding of the glycerol electro-oxidation reaction and help to take steps forward on the development of fuel cells and electrolyzers fed by this alcohol. One of the first steps was given by Lamy's group in the 1990s who investigated the GEOR on a well-ordered Au surface (Avramov-Ivić et al. 1991a) and identified some products by using high-performance liquid chromatography (HPLC) (Roquet et al. 1994b). This work evidenced the importance of the catalytic surface on the conversion of glycerol in current density and the importance of the experimental medium to produce different carbonyl compounds via electrochemical reactions.

The analysis of the products of GEOR is feasible by using HPLC to separate them and a UV-based detector to identify them. The electrochemical reaction generates an ultra-thin layer of substances close to the electrode surface, and relatively long times of reaction are needed to accumulate an amount of compounds detectable by HPLC. Therefore, long time electrolysis is performed to increase the amount of products in solution, while aliquots are removed and injected on the chromatograph. This kind of ex situ experiment is quite important to mimic real systems, but multiple steps are required for the analysis. Kwon and Koper developed a breakthrough technique for analyzing products of GEOR and other electrochemical reactions (Kwon and Koper 2010). These authors combined an electrochemical surface reaction with HPLC by online sampling solution close to the electrode surface during electrolysis. This setup allows fast sampling from the solution precisely close to the electrode surface during an electrochemical experiment. Online HPLC coupled to electrochemical experiments, traditional HPLC, and others experiments based on these have been used to investigate products of GEOR on different catalysts (Soares et al. 2016; Thia et al. 2016; Da Silva et al. 2017).

Spectroscopy techniques have also been used to identify products and guideline the pathways of GEOR. The classic and powerful in situ FTIR has been used by

Camara's and Tremiliosi-Filhos's groups and coworkers for study such reaction (Martins et al. 2011; Fernández et al. 2012b; Gomes et al. 2012). In this configuration, a spectroelectrochemical cell is used to allow the interaction between the infrared beam and a thin layer of substances formed between a working electrode and a bottom window (transparent for infrared) during the course of the reaction. This configuration allows the researcher to follow in situ the formation of intermediates and products, whose nature is inferred by the bands that emerge in a spectrum. Details and examples of in situ FTIR are found elsewhere (Iwasita and Nart 1997). Moreover, NMR spectroscopy has been recently used to identify products of GEOR (Fashedemi et al. 2015; Huang et al. 2016). Although very promising, this last technique is not subject of this chapter.

Therefore, both HPLC-based techniques and in situ FTIR are powerful tools to help understanding the pathways and mechanisms, in some extent, of GEOR. The pathways of the reaction are strongly influenced by the catalysts, including the type of metal, e.g., Au, Pt and their alloys, and the atomic surface arrangement. GEOR is so sensitive to the atomic surface arrangement that changing the crystallography of Pt catalyst from a specific configuration [i.e., (111)] to a more less energetic [i.e., (100)] changes the product (Garcia et al. 2016). In this sense, it is fundamental to keep researches focused on finding the best catalyst structure to enhance activity in terms of energy conversion and in terms of selectivity of the reaction. On the other hand, real systems require the use of nanomaterials as anodes to enhance the surface to volume ratio. In this sense, the findings of well-ordered surfaces must be adapted to nanoscale. This issue will be addressed in item 2.2.

There are plenty of methods to synthesize NPs, since well-ordered (Hutchings 2008; Jin et al. 2011; Chen et al. 2015) to multi-metallic nanomaterials (Vidal-Iglesias et al. 2010; Ding et al. 2016; Alencar et al. 2016; Caneppele et al. 2017), whose discussion is out of the scope of this chapter. The most important point here is to clarify how important is the chemical composition of the NPs on the activity and selectivity of GEOR. In the following items, we show some proposed pathways of GEOR on Pt and Au surfaces described by us and other groups, as a result of chromatographic and spectroelectrochemical measurements. Moreover, we illustrate how the atomic surface arrangement and chemical composition might be designed to improve activity and enhance selectivity.

## 7.2.1 Proposed Mechanisms for Glycerol Electro-Oxidation

Back to middle 1990s, Lamy's group was the precursor team of researchers to unravel some of the pathways involved in the electro-oxidation of glycerol (Roquet et al. 1994a). By using HPLC, the authors collected aliquots of acid and alkaline solution in which glycerol was electrolyzed over a polycrystalline smooth platinum surface in long-term experiments (around 30 h). Authors realized that some parameters influence the nature of products and intermediates, as the concentration of the reactant, the applied potential, and the pH of the medium (Roquet et al. 1994a). In this study, several substances were detected by HPLC. Namely, the production of glyceraldehyde, tartronic acid, glyceric acid, glycolic acid, formic acid, and oxalic acid was confirmed (the last three substances are formed when the carbon chain of glycerol is broken) (Roquet et al. 1994a). In that occasion, the authors explained the presence of such oxidation products in terms of interactions between glycerol molecules (or their intermediates) and the surface of platinum. Although the explicit mechanism is too technical (and therefore beyond the scope of this text), a general framework can be traced, based on the original work.

Before detailing the pathways of GEOR, we will discuss some works that have shed a light into the mechanism of adsorption and the first steps of the reaction (Martins et al. 2011; Gomes et al. 2013; Garcia et al. 2016). The desired product for energy conversion application is $CO_2$, which means the glycerol was totally electro-oxidized. In this context, Martins et al. (2011) investigated the routes to total oxidation by using in situ FTIR. These authors integrated the bands taken from the FTIR experiments (Fernández et al. 2012b) to make it comparable to the electrochemical measurements. The authors found that CO is produced at early stages, at low potentials, with a maximum that coincides with the beginning of $CO_2$ production. Hence, it became clear that glycerol experiences dissociative adsorption on Pt surfaces forming CO, which is oxidized to $CO_2$. However, the formation of $CO_2$ kept increasing at high potentials after the observable consumption of CO, suggesting that other pathway(s) must be involved. The authors found that carbonyl compounds (mainly glyceric acid) appeared at the same time as $CO_2$ and are consumed while $CO_2$ production increases (Martins et al. 2011). In this sense, these authors suggested a mechanism (mechanism here is defined as a pathway considering the contribution of the surface) for GEOR. At low potentials, Pt in acid medium is available for glycerol adsorption, producing CO, which generates $CO_2$. The increase in potential promotes the conversion of glycerol to partially oxidized carbonyl compounds. Concurrently, an oxide layer made of –O and –OH grows on Pt surface while the potential is increased. Hence, at high potentials, those carbonyl compounds might react with –O and –OH groups from Pt surface to form $CO_2$ (Martins et al. 2011). The adsorption of glycerol on Pt was investigated by the same research group (Gomes et al. 2013). The authors found that glycerol molecules compete with H species by Pt active sites. Hence, the CO formation is dependent on the H desorption step (Gomes et al. 2013).

By using online mass electrochemical spectroscopy, Garcia et al. (2016) investigated the adsorption steps and mechanism of GEOR. These authors found that the reaction might follow two main mechanisms depending on the atomic surface arrangement of the Pt surface, as discussed in the item 2.2.1. Here, we would like to highlight the existence of two main adsorbed intermediates, one double bonded to Pt by the terminal carbon and another one single bonded to Pt by the terminal and central carbons and each one leads to different pathways (Garcia et al. 2016). In an attempt to understand the contribution of each of the three carbon on the $CO_2$ production, Fernández et al. (2012b, 2013) followed the reaction by in situ FTIR using isotopically labeled glycerol. As expected, the authors found the terminal carbons are more likely to be oxidized, mainly due to an steric probability, but more interestingly, they found that $CO_2$ production from terminal carbon is 19 times more intense than from the central one (Fernández et al. 2012b).

There are a plenty of publications proposing new routes, pathways, and mechanisms for GEOR. The real mechanism is really far from full understanding, and the few works regarding the steps of the reaction are those responsible for elucidating it. Most of the pathways published are highly useful to shed a light on the possible products but do not help to clarify the whole mechanism. In our point of view, the mechanism must consider the presence of the catalyst and adsorbed species accordingly. Figure 7.5 gives a small idea of such complex possible pathways that glycerol might undergo during its electro-oxidation, considering the publications to date.

Figure 7.5 shows an overview of the possible pathways that might occur in both acid and alkaline medium. The only difference would be the protonated or non-protonated acid hydrogen at the molecules. Moreover, in alkaline medium, the production of carbonate ($CO_3^{2-}$) is expected instead of $CO_2$ in the bulk solution. The trigger of glycerol electro-oxidation might be summarized in two paths, one started with dihydroxyacetone and another one initiated with glyceraldehyde.

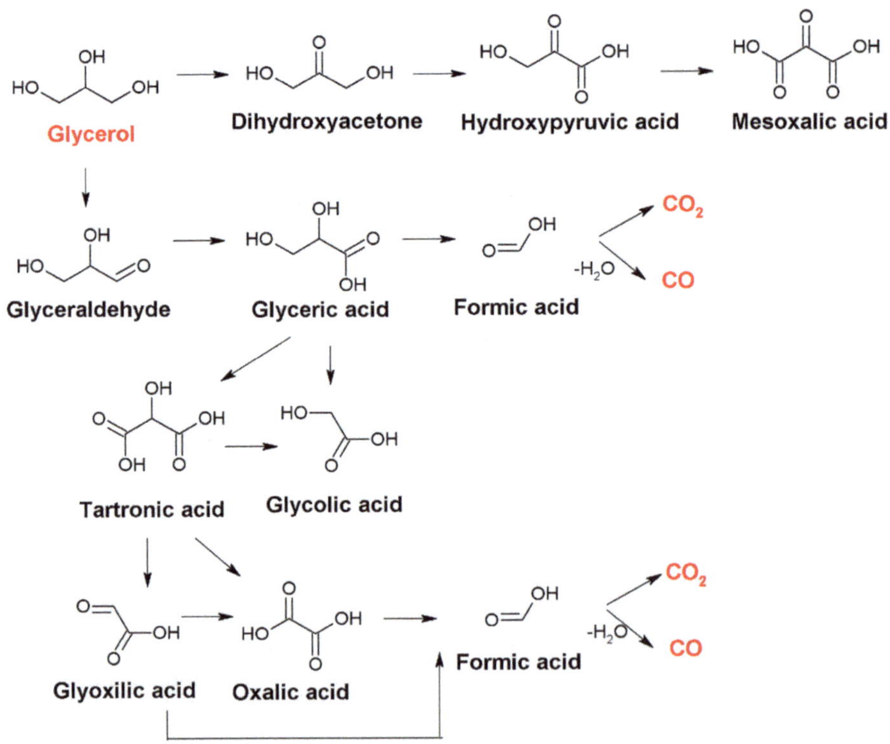


**Fig. 7.5** Illustrative suggested pathways for glycerol electro-oxidation reaction. It has been reported similar pathways for both acid and alkaline medium, with the difference at the protonated acid hydrogen

The "golden" route for the conversion of glycerol into high valuable compounds is the first one, since hydroxypyruvic acid is the most desirable compound (see prices in Table 7.1). Moreover, dihydroxyacetone has elevated market price and several industrial applications. The second route also leads the reaction toward valuable carbonyl compounds with the additional path toward total electro-oxidation, which allows concomitant production of energy and fine chemicals. Worth noticing that, regardless the selective multi-metallic catalysts used as anode for GEOR to date, all reports show pathways that might be found in Fig. 7.5. The well-ordered, multi-metallic, bulk, or nanomaterials work to increase the amount of one or another byproduct.

## 7.2.2 Catalysts: Bulk and Nanomaterials

The complex mechanisms and pathways proposed for GEOR on Pt and Au surfaces might change for other catalytic surfaces. Moreover, the use of a metal in nanoscale might modify the pathways of GEOR compared to the bulk catalyst, which is commonly called the *nano effect*. Such effect is still very difficult to prove experimentally, since the classic half-cell measurements performed in electrochemistry show an average response of the whole catalyst; namely, it is the feature of a summation of all NPs immobilized on an electron collector (working electrode). Although much effort have been spent to investigate individual electrochemical responses of single NPs (Shan et al. 2012; Fang et al. 2014; Yu et al. 2015), there is no work regarding the investigation of a complex reaction such as GEOR on single NPs. In this sense, although the use of spectroelectrochemical and chromatographic techniques deeply helps to understand the pathways, the real contribution of the NPs on the GEOR is still unknown to date.

In the next sections, we show and discuss published results of our group and others regarding the influence of the surface atomic arrangement and the surface chemical composition of catalysts on the glycerol electro-oxidation pathways.

### 7.2.2.1 The Influence of the Surface Arrangement on the Activity and Selectivity of Glycerol Electro-oxidation

The complexity of a real catalyst was commented in Sect. 7.1.3. Experiments using single crystals are very useful to understand how the atoms are distributed over the surface of a catalyst and how such arrangement influences the selectivity and rate of a given reaction. On this sense, by using a surface which we "know" the positions of each atom may be of great help to clarify their role in catalysis (and electrocatalysis). This can be achieved if the atoms are organized over a surface. In this way, the position of each particular atom will obey to a simple-periodic function and any atom will be representative of the whole surface (with few exceptions corresponding to the edges of the sample). Researchers can design bulk materials where virtually all the atoms at the surface have the same configuration; they are called “single crystals.” For instance, two of the so-called basal planes structures are shown in Fig. 7.6. A polycrystalline surface is one where the atoms are arranged in different ways in different parts of the surface. In a polycrystalline material, we have regions having atoms with (100) configurations, other with (111), other with (110), steps, kinks, etc. Summarizing, it is a generally rough inhomogeneous surface and all the metals in nature present this kind of arrangement. In contrast, in a single crystal structure like those shown in Fig. 7.4, the atoms are arranged forming very specific structures. We invite the readers to compare the configurations of the atoms in Fig. 7.6 with those of the NP in Fig. 7.4. These arrangements might be found on the surface of NPs if they were designed to be well-shaped, such as nanocubes, nanoctahedron, and others (Solla-Gullón et al. 2013).

The GEOR was studied in acid media using the Pt basal planes as well as disordered single crystals (Gomes et al. 2012; Fernández et al. 2015b, 2016). Even if the atoms arrangements strongly influence the catalytic activity of the material, the reaction occurs at very low rates below 0.6 V for any configuration. The lowest overpotential is obtained with Pt(111), and highest currents are obtained with Pt (100) well-ordered or defected. Recently, important aspects regarding the selectivity of this reaction were addressed by using sample collection on line (Garcia et al. 2016). Considering those products that are formed without breaking of C–C bonds (i.e., products of intact chain), it was shown that both the secondary and primary carbon of glycerol are oxidized on Pt(111). On the other hand, the reaction is selective toward the oxidation of the terminal carbons on Pt(100). The results were explained in terms of a different configuration of possible intermediates of the reaction on Pt(111) and Pt(100). The GEOR was not studied on another single crystal surface apart from Au(100) in alkaline media as Au is not active for the oxidation of small organic molecules in acid media (Avramov-Ivić et al. 1991b).

These fundamental results provide important practical information, since well-shaped Pt NPs might be desired to contain extended (100) and (111) surfaces

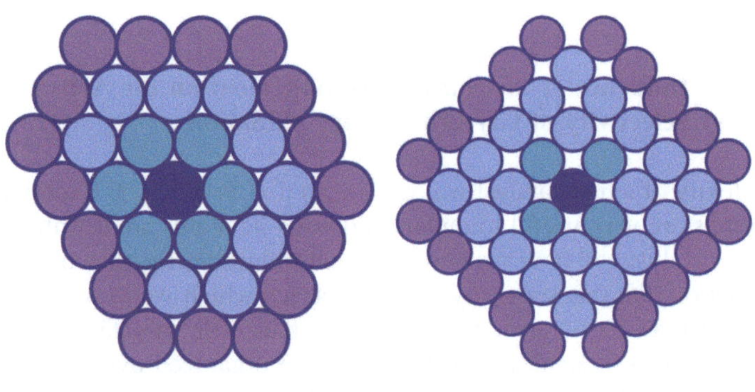

**Fig. 7.6** Illustration of well-ordered (111) and (100) surfaces. In the (111) surface, each atom is coordinated to six atoms in the same plane. In the (100), the atoms are coordinated to four atoms in the same plane

in attempt to present high current densities and low onset overpotential, respectively. In this context, to bridge the gap between model and applied systems, it is interesting to perform experiments using well-shaped nanoparticles. An example of a well-shaped NP is that already shown in Fig. 7.4. There are some methods which permit to obtain NPs with structures similar to those shown in Fig. 7.7 (Solla-Gullón et al. 2013). In a cubic NP, most of the atoms have the configuration of those in a (100) single crystal. On the other hand, an octahedral NP maximizes those atoms having a (111) configuration. Even if these are model systems, because we have more reliable information of their structure compared to "real" catalysts, they contain intrinsic defects and they can be incorporated on different supports. Thus, after a detailed study using single crystals, we can go a step closer to real system using these well-shaped materials.

There are no papers regarding the GEOR on well-shaped Pt nanoparticles; however, the experiments performed using single crystals and randomly defected single crystals, which can mimic the behavior of well-shaped NPs, would permit to predict the results (Fernández et al. 2015a, b, 2016). The reaction was studied in alkaline media on Au cubic and octahedral NPs and on Pd cubic and spherical NPs by Monzó et al. (2015). The onset of the reaction on Au NPs is similar for cubic and octahedral NPs; however, octahedral NPs are more active, which could be related to

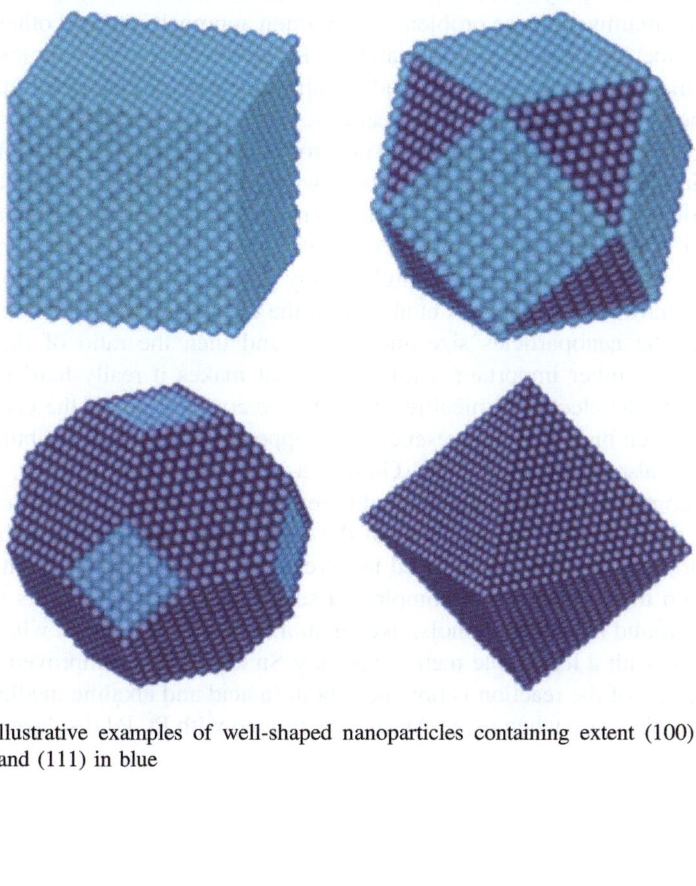

**Fig. 7.7** Illustrative examples of well-shaped nanoparticles containing extent (100) surface in light blue and (111) in blue

a higher activity of (111) domains. Pd nanocubes are more active than nanospheres, which are probably connected with a higher activity of (100) domains. However, to the best of our knowledge, there is no study of the GEOR using the three single crystals basal planes of Pd and Au, which would permit to confirm if the activity trends correlate with the relative quantity of defects, ordered domains, and their atoms configuration.

#### 7.2.2.2 The Influence of the Chemical Composition of the Catalyst on the Activity and Selectivity of Glycerol Electro-oxidation

Pt has been extensively used for the electro-oxidation of small organic molecules in acid and alkaline media. However, regardless the surface atoms arrangement, the catalyst suffers of poisoning due to the strong adsorption of partially oxidized molecules as, for example, CO (Gomes et al. 2013). Gold has also been extensively used for the oxidation of organic molecules in alkaline media. In this case, the electrode does not suffer from poisoning; however, the reaction onset potentials are higher than that for Pt, which means that we need either to apply more energy to produce the reaction (electrolyzer) or that the anode of the fuel cell will work at higher potentials, decreasing the cell voltage and the available power. Besides, both materials in their polycrystalline form are non-selective to any product of the GEOR. To circumvent these problems, a common approach is to add other atoms to pure noble metal materials. Unfortunately, it highly increases the complexity of the materials making difficult to understand the physicochemical parameters that indeed modify the activity/selectivity of the catalyst.

There are many papers about the electro-oxidation of glycerol using bi and tri-metallic nanoparticles. Apart from those where the second (or second and third) metal is deposited on a pre-existing nanoparticle (generally called "nanoparticle decoration"), the introduction of other metals highly affects the physicochemical properties of the material. Even in high quality papers where authors produce NPs in highly controlled setups (Park et al. 2008), the changes in the composition of the NPs affect the nanoparticles size and shape (and then the ratio of the exposed facets), among other important parameters, what makes it really hard to connect differences in the electrochemical results with the composition of the catalyst. The problem is even bigger in the presence of a support, because the distribution of the NPs in it is also severely affected (Gomes et al. 2014). Unfortunately, it is even harder to compare results obtained by different authors, where we have commonly important differences in the methods of the synthesis of the materials, the kind of the support, and in the methods used to investigate the activity/selectivity, etc.

Although these systems are complex, it seems that glycerol follows the trends previously found for other alcohols, like ethanol and methanol. Thus, when Pt or Pd is combined with a less noble metal (typically Sn and Ru), an improvement in the onset potential of the reaction is obtained (both in acid and alkaline media). Similar features are observed when using Au, which alloyed with Pt, Pd displaces the onset of the reaction closer to the thermodynamic reversible potential (Lertthahan et al. 2017).

In our opinion, up to now, it is almost impossible to reach an agreement about current densities (activity) and especially about how the selectivity is affected by different multi-metallic catalysts. In consequence, we will restrict ourselves to review some few works using ad-atoms, i.e., atoms deposited in a controlled way on noble bulk and NPs metals.

The effect of many p-block metals on polycrystalline (Kwon et al. 2014) and single crystal (Garcia et al. 2017) electrodes in acid media was studied by Koper's group. Bi, Sb, Pb, Sn, and In reversibly or irreversibly adsorbed increase the Pt activity toward the GEOR. Both Sb and Bi promote the oxidation selectively to dihydroxyacetone (DHA) (Kwon et al. 2014). Irreversibly adsorbed Bi on Pt(111) increases the surface activity and selectivity toward DHA. On the other hand, the modification of Pt(100) decreases the activity without changing the selectivity, i.e., Pt(100) with and without Bi selectively oxidizes the terminal carbon generating glyceraldehyde and glyceric acid (Garcia et al. 2017).

The very good results obtained using Sb encouraged Cannepele et al. (2017) to study how the activity of polycrystalline Pt NPs is affected by the coverage of Sb. The authors verified that the electrode activity augments exponentially with the coverage until $\theta_{sb} = 0.81$, which is the maximum stable coverage, at least under the reported experimental conditions. Finally, very interesting results were obtained by Zalineeva et al. (2013) using cubic and spherical Pd NPs. While the decoration of the spheres with Bi diminishes the activity of the electrode, the opposite effect is observed for the cubes.

## 7.3 Fuel Cells and Electrolyzers Fed by Glycerol

After much effort to develop new anodes for glycerol electro-oxidation, some researchers have been working in engineering real systems to use clean and even crude glycerol, obtained directly from biodiesel fabrication. Promising results were reached to produce clean hydrogen gas through glycerol electrochemical reforming (Marshall and Haverkamp 2008; González-Cobos et al. 2016). Marshall and Haverkamp used a proton exchange membrane electrolyzer fed by glycerol on a Pt immobilized on Ru-Ir oxides (Marshall and Haverkamp 2008). These authors produced 10 $m^3$ of $H_2$ per day per $m^3$ of the reactor, consuming 1.1 kWh per $m^3$ of $H_2$, which is a remarkable result since it is 66% less energy-consuming than water electrolysis (Marshall and Haverkamp 2008). González-Cobos et al. (2016) used PtBi/C catalyst for glycerol electrolysis to improve the pure $H_2$ production, which was 57–66% less energy-consuming than water electrolysis.

There are several reports about fundamental investigations using well-cleaned and $O_2$-free solution during half-cell measurements for glycerol electrolysis, but there is still a lack of current reports considering configurations closer to real systems, suggesting proper engineering. Among the few reports, we might cite

Kongjao et al. (2011) who used a 0.5 L cylinder-shape reactor containing two Pt grids at constant 400 rpm stirring for glycerol electrolysis. Using chromatographic methods, the authors found propanediol, glycidol, and 2-propenol after 6 h of glycerol electrolysis in acid medium under 4.5 A of galvanostatic experiment (Kongjao et al. 2011). Yuksel et al. (2010) used a flow reactor for glycerol electrolysis on titanium electrodes (as anode and cathode). The authors reported full description of the system, including furnace, heater, pump, heat exchanger, back pressure regulator, and DC supplier. The electrolysis was carried out in alkaline solution, and the authors found $H_2$, glyceraldehyde, glycolaldehyde, lactic acid, formic acid, glycolic acid, and acetic acid as main products (Yuksel et al. 2010). Still regarding more practical devices, Lam et al. published a breakthrough system using Pt-free catalyst, by using low loading cobalt-based anodes for glycerol electrolysis in a stirring system (Lam et al. 2017). The authors tailored the products by changing temperature and alkalinity of the medium. They detailed how to use atmospheric pressure to electrosynthesize lactic acid at current densities close to 1.8 mA $cm^{-2}$ at 60 °C while glyceric acid is produced at room temperature and 8.8 mA $cm^{-2}$ (Lam et al. 2017).

Although, to the best of our knowledge, there is no commercial systems or companies producing energy and high valuables compounds concomitantly, there are initial investigation and devices at laboratorial scale regarding energy production using fuel cells fed by glycerol (Roquet et al. 1994b; Sau and Rogach 2010; Kwon and Koper 2010; Martins et al. 2011, 2014a; Fernández et al. 2012b; Gomes et al. 2012; Horn et al. 2016; Soares et al. 2016; Coutanceau and Baranton 2016; Thia et al. 2016; Da Silva et al. 2017). These works might be gathered as (i) solid oxide fuel cells (SOFC) (Won et al. 2009; Lima da Silva and Müller 2010), (ii) Au-based anodes (Zhang et al. 2012b; Xin et al. 2012), (iii) Pd-based anodes (Marchionni et al. 2013; Benipal et al. 2017a, b), (iv) Pt-based anodes (Matsuoka et al. 2005; Ilie et al. 2011; Zhang et al. 2012c; Qi et al. 2013; Wang et al. 2014), and (v) glycerol microfluidic fuel cells (Dector et al. 2013; Maya-Cornejo et al. 2016).

Won et al. (2009) used tubular SOFC fed by glycerol with internal reforming. The authors found 265 mW $cm^{-2}$ as maximum power when the fuel cell was operated at 800 °C. Moreover, at 700 °C, the SOFC was stable for more than 100 h (Won et al. 2009). A full thermodynamics of SOFC fed by glycerol can be found elsewhere (Lima da Silva and Müller 2010). Zhang et al. built an anion-exchange-membrane direct glycerol fuel cell (AEM-DGFC) more reactive than those fed by methanol and ethylene glycol in alkaline medium (Zhang et al. 2012b). Although a Fe-based catalyst was used as cathode, the good performance was rationalized as a consequence of the use of small amount of Au/C at the anode. At 80 °C, the open circuit voltage (OCV) reached 0.67 V and the power density 57.9 mW $cm^{-2}$. Using crude glycerol, the authors found 30.7 mW $cm^{-2}$ at 0.66 V of OCV. The authors also identified some partially oxidized compounds, as tartronic acid, mesoxalic acid, and oxalic acid (Zhang et al. 2012b). Still using small Au/C NPs and Fe-based catalyst as cathode, Xin et al. (2012) suggested the cogeneration of mesoxalic acid and electricity at the anode of a AEM-DGFC.

Benipal et al. (2017b) used PdAg NPs dispersed on carbon nanotubes as anodes for 2.0 M KOH + 1.0 M glycerol at 2.0 mL $min^{-1}$ in an AME-DGFC to produce an OCV of 0.83 V and 76.5 mW $cm^{-2}$ of power density. After constant applied potential, the authors found glycerate, tartronate, mesoxalate, lactate, glycolate, and oxalate as products. Using similar materials, the same group reported remarkably 214.7 mW $cm^{-2}$ of power density at 80 °C by using polytetrafluoroethylene as anode/cathode separator (Benipal et al. 2017a). Marchionni and coauthors used Pd-(Ni-Zn)/C anodes in DGFC (Marchionni et al. 2013). These authors fed an anode with 2.0 M KOH + 5 wt% glycerol at 80 °C to produce 119 mW $cm^{-2}$. According to the authors, such increase in power density compared to Pd/C NPs (77 mW $cm^{-2}$) suggests fast kinetics due to the non-alloyed structure of the catalyst (Marchionni et al. 2013).

Even still scarce, the most extensive amount of reports regarding DGFC show Pt-based anodes. Matsuoka et al. compared the performances of DAFC fed by different alcohols using Pt–Ru/C as anode and Pt/C as cathode in alkaline medium at ~50 °C (Matsuoka et al. 2005). The maximum power densities increase in the following order: xylitol, erythritol, methanol, glycerol, and ethylene glycol. The authors reported power density of ~6 mW $cm^{-2}$ by feeding the anode with glycerol in such configuration (Matsuoka et al. 2005). Wang et al. (2014) reported 86 mW $cm^{-2}$ and 0.73 V as maximum power density and OCV for AEM-DGFC with Pt/C as anode fed by pure glycerol and Ag NPs as cathode for oxygen reduction reaction. Using Pt/C as anode, Ilie et al. (2011) found power density in the order of 10 mW $cm^{-2}$, which was doubled only by changing parameters as the membrane exchange assembly fabrication.

Outstanding results were achieved by Qi et al. (2013) by using surface dealloyed Pt–Co dispersed on carbon nanotubes anode fed by 6.0 M KOH + 3.0 M glycerol (4.0 mL $min^{-1}$) while $O_2$ was reduced on Fe–Cu–$N_4$/C cathode. The authors reported 268.5 mW $cm^{-2}$ as maximum power density, which is still the best performance for low-temperature single glycerol fuel cell (Qi et al. 2013), comparable to a SOFC fed by glycerol (Won et al. 2009). Regarding membrane exchange assembly (MEA) cells, there is no doubt that alkaline medium is more promising for anodes fed by glycerol than acid medium to produce high power. Moreover, as the DGFC was not fully explored, Au-, Pd- and Pt-based are still considered potential anodes for fuel cells and electrolyzers.

Another breakthrough technology for low-power devices as portable and wearable electronics and sensing is the microfluidic fuel cells (MFCs). Briefly, these types of fuel cells do not require ionic exchange membrane, which decrease their prices and loss of potential. Furthermore, the absence of MEA allows the use of previous non-compatible media, as alkaline medium as anolyte with acid medium as catholyte. A complete description can be accessed elsewhere in book (Kjeang 2014) and reviews (Wang et al. 2016; Goulet and Kjeang 2014). Such recent technology showed itself very encouraging for glycerol powering (Dector et al. 2013; Maya-Cornejo et al. 2016). Dector et al. 2013 investigated the performance of a Y-type microfluidic fuel cell fed by $O_2$-free 0.1 M glycerol as anolyte and $O_2$-saturated 0.3 M KOH as catholyte, using Pt/C as cathode and Pd/C or Pd dispersed on multi-walled carbon nanotubes (MWCNTs) as anode. In this pioneer work, power in the order of 0.7 mW $cm^{-2}$ was achieved. Another work from the same group showed an air-breathing cathode nanofluidic flow-through fuel cell fed by pure and crude glycerol (Maya-Cornejo et al. 2016). In that work, Maya-Cornejo and coworkers dispersed Cu@Pt/C NPs on carbon nanofoam and used as anode and an air-breathing Pt/C cathode, generating 23 and 21.8 mW $cm^{-2}$ for pure and crude glycerol, respectively. Therefore, the scalability of MFC might be more explored for low-powered devices or even ordinary electronic devices.

## 7.4 Perspectives and Future Engagements

Considering that the massive fabrication of biodiesel as an economic and environmental friendly matrix for countries around the world is relatively recent, the advance in the field of glycerol in electrochemistry is impressive. As it happens for new technologies, the bottleneck for scalable application of glycerol for energy, hydrogen, or chemical production is the industrial gap. There is a barrier between fundamental laboratorial results and real applications, which needs to be overcome. Experimental and theoretical researches for fundamental and applied science are still required. Here, we point out our outlook for future indispensable researches.

From the fundamental point of view, new stable well-ordered multi-metallic nanomaterials must be synthesized and investigated. The ordered surface structure summed to the bimetallic (at least) composition is the best option to reach selective and active catalysts for such complex reaction. On the other hand, these catalysts must be stable. Although the stability investigation of a nanomaterial is hardly publishable by itself, it is greatly important for the application point of view and, sometimes, more important than a once-in-life good catalyst. In this sense, researchers and industrials need to compare results among papers, reports, and patents, which are very difficult to achieve since the scientific community in this field has no standard protocol to evaluate any of the parameters previously described in this chapter.

Hence, the development of consensual protocols is imperative. The most important point to allow comparison among different works is to detail the experimental procedure, including medium, concentration of electrolyte and glycerol, temperature, potential range, loading of catalysts, flow rate. We highly recommend the use of reversible hydrogen electrode (RHE) as reference potential scale. Even a simple comparison, as it is the potential range, is very difficult considering that some researchers do not present details about the medium and there is no Ohmic drop compensation in most of them (even when needed), which makes very hard the re-scaling. In terms of current density, we recommend to clearly show the method used to obtain the electrochemically surface area (when necessary) for both half-cell and single fuel cell measurements. Moreover, it is interesting for industrials if a researcher shows a specific activity in terms of current normalized by total mass (or partial mass, depending on the case) of the catalyst, and such information gives an idea of cost-effectiveness and scalability.

In our point of view, the most critical parameter concerns the stability of a catalyst. All methods, i.e., following active surface area during cycle, current density during cycles, microscopy images during cycles, or even following pseudo-current density at constant potential, are very useful, but only if the results were obtained from comparable experimental conditions. A very useful tool for stability tests is to use synchrotron-based techniques. There are several synchrotron facilities around the world that can be used for any researcher after acceptance of a research proposal. These machines permit to study the structure of many materials (and many other physicochemical parameters) during a given electrochemical experiment.

Regarding the production of high valuable compounds, we ponder that it is mistakable to sell an idea of cogeneration of chemical and electricity if the sampling process is taking place in a different step (i.e., the authors perform polarization curves to record power in a set of experiments and long-time electrolysis in another set of experiments). In this case, there is just a potential chance of dual process, since power is obtained in an instrument while high valuable compounds are usually obtained in another one. For these reports, the idea of cogeneration might induce the reader to a misunderstanding. Moreover, the comparison of the selectivity of different catalysts is very difficult, since the works report different experimental conditions, as applied potential, medium, temperatures, etc.

Other unexplored area is the separation of glycerol-derived compounds. Until this moment, there is only chromatographic separation performed in high-performance liquid (or gas) chromatography equipment, which uses small and high pressure columns with complex special stationary phase. However, the stability of such columns must be investigated for further scalable production. Most important, new preparative liquid chromatography is pivotal to get glycerol electrolysis in an industrial plant level. In attempt to make it real, researchers from fundamental areas must work together with engineers, who can use the recent efforts to design real systems. Here, we might cite the need (i) for higher level of glycerol utilization for both fuel cell and electrolysis, (ii) for a less stepped process, and (iii) for automation.

The advances in glycerol electrochemical technologies are so encouraging that have recently risen up the number of publications and patents, so specific economic studies of viability and applicability are needed. The use of glycerol for electrochemical purposes proves itself an interesting smart way to use this byproduct of biodiesel fabrication. We hope to find these environmental friendly technologies in use soon.

**Acknowledgements** Authors acknowledge Renzzo M. Giz for illustrations in Figs. 7.4, 7.6, and 7.7. The authors acknowledge financial assistance from Conselho Nacional de Desenvolvimento Científico e Tecnológico (CNPq, Grant # 454516/2014-2), Fundação de Apoio ao Desenvolvimento do Ensino, Ciência e Tecnologia do Estado de Mato Grosso do Sul (FUNDECT, Grants # 026/2015 and #099/2016), Coordenação de Aperfeiçoamento de Pessoal de Nível

Superior (CAPES), Fundação de Amparo à Pesquisa do Estado de São Paulo (FAPESP) and Finaciadora de Estudos e Projetos (FINEP).

## References

Alencar LM, Caneppele GL, Martins CA (2016) Fast and controlled decorating of metallic nanoparticles using wall-jet configuration. Electroanalysis 29:314–318. https://doi.org/10.1002/elan.201600472

Avramov-Ivić ML, Leger JM, Lamy C et al (1991a) The electro-oxidation of glycerol on the gold (100)-oriented single-crystal surface and poly crystalline surface in 0.1 M NaOH. J Electroanal Chem Interfacial Electrochem 308:309–317. https://doi.org/10.1016/0022-0728(91)85075-Z

Avramov-Ivić ML, Leger JM, Lamy C et al (1991b) The electro-oxidation of glycerol on the gold (100)-oriented single-crystal surface and poly crystalline surface in 0.1 M NaOH. J Electroanal Chem Interfacial Electrochem 308:309–317. https://doi.org/10.1016/0022-0728(91)85075-Z

Bagnato G, Iulianelli A, Sanna A, Basile A (2017) Glycerol production and transformation: a critical review with particular emphasis on glycerol reforming reaction for producing hydrogen in conventional and membrane reactors. Membranes 7:17. https://doi.org/10.3390/membranes7020017

Bambagioni V, Bevilacqua M, Bianchini C et al (2010) Self-sustainable production of hydrogen, chemicals, and energy from renewable alcohols by electrocatalysis. Chemsuschem 3:851–855. https://doi.org/10.1002/cssc.201000103

Babir F (2012) PEM fuel cells: theory and practice, 2nd edn.

Benipal N, Qi J, Gentile JC, Li W (2017a) Direct glycerol fuel cell with polytetrafluoroethylene (PTFE) thin film separator. Renew Energy 105:647–655. https://doi.org/10.1016/j.renene.2016.12.028

Benipal N, Qi J, Liu Q, Li W (2017b) Carbon nanotube supported PdAg nanoparticles for electrocatalytic oxidation of glycerol in anion exchange membrane fuel cells. Appl Catal B Environ 210:121–130. https://doi.org/10.1016/j.apcatb.2017.02.082

Bianchini C, Shen PK (2009) Palladium-based electrocatalysts for alcohol oxidation in half cells and in direct alcohol fuel cells. Chem Rev 109:4183–4206. https://doi.org/10.1021/cr9000995

Bott-Neto JL, Garcia AC, Oliveira VL et al (2014) Au/C catalysts prepared by a green method towards C3 alcohol electrooxidation: a cyclic voltammetry and in situ FTIR spectroscopy study. J Electroanal Chem 735:57–62. https://doi.org/10.1016/j.jelechem.2014.10.010

Caneppele GL, Almeida TS, Zanata CR et al (2017) Exponential improving in the activity of Pt/C nanoparticles towards glycerol electrooxidation by Sb ad-atoms deposition. Appl Catal B Environ 200:114–120. https://doi.org/10.1016/j.apcatb.2016.06.072

Chen Q-S, Xu Z-N, Peng S-Y et al (2015) One-step electrochemical synthesis of preferentially oriented (111) Pd nanocrystals supported on graphene nanoplatelets for formic acid electrooxidation. J Power Sour 282:471–478. https://doi.org/10.1016/j.jpowsour.2015.02.042

Coutanceau C, Baranton S (2016) Electrochemical conversion of alcohols for hydrogen production: a short overview. Wiley Interdiscip Rev Energy Environ 5:388–400. https://doi.org/10.1002/wene.193

Da Silva RG, Aquino Neto S, Kokoh KB, De Andrade AR (2017) Electroconversion of glycerol in alkaline medium: from generation of energy to formation of value-added products. J Power Sources 351:174–182. https://doi.org/10.1016/j.jpowsour.2017.03.101

Dector A, Cuevas-Muñiz FM, Guerra-Balcázar M et al (2013) Glycerol oxidation in a microfluidic fuel cell using Pd/C and Pd/MWCNT anodes electrodes. Int J Hydrog Energy 38:12617–12622. https://doi.org/10.1016/j.ijhydene.2012.12.030

Ding J, Bu L, Guo S et al (2016) Morphology and phase controlled construction of Pt–Ni nanostructures for efficient electrocatalysis. Nano Lett. https://doi.org/10.1021/acs.nanolett.6b00471

Falase A, Main M, Garcia K et al (2012) Electrooxidation of ethylene glycol and glycerol by platinum-based binary and ternary nano-structured catalysts. Electrochim Acta 66:295–301. https://doi.org/10.1016/j.electacta.2012.01.096

Fang Y, Wang W, Wo X et al (2014) Plasmonic imaging of electrochemical oxidation of single nanoparticles. J Am Chem Soc 136:12584–12587. https://doi.org/10.1021/ja507097y

Fashedemi OO, Miller HA, Marchionni A et al (2015) Electro-oxidation of ethylene glycol and glycerol at palladium-decorated FeCo@Fe core–shell nanocatalysts for alkaline direct alcohol fuel cells: functionalized MWCNT supports and impact on product selectivity. J Mater Chem A 3:7145–7156. https://doi.org/10.1039/C5TA00076A

Fernández PS, Fernandes Gomes J, Angelucci CA et al (2015a) Establishing a link between well-ordered Pt(100) surfaces and real systems: how do random superficial defects influence the electro-oxidation of glycerol? ACS Catal 5:4227–4236. https://doi.org/10.1021/acscatal.5b00451

Fernández PS, Martins CA, Angelucci CA et al (2015b) Evidence for independent glycerol electrooxidation behavior on different ordered domains of polycrystalline platinum. ChemElectroChem 2:263–268. https://doi.org/10.1002/celc.201402291

Fernández PS, Martins CA, Martins ME, Camara GA (2013) Electrooxidation of glycerol on platinum nanoparticles: deciphering how the position of each carbon affects the oxidation pathways. Electrochim Acta 112:686–691. https://doi.org/10.1016/j.electacta.2013.09.032

Fernández PS, Martins ME, Camara GA (2012a) New insights about the electro-oxidation of glycerol on platinum nanoparticles supported on multi-walled carbon nanotubes. Electrochim Acta 66:180–187. https://doi.org/10.1016/j.electacta.2012.01.069

Fernández PS, Martins ME, Martins CA, Camara GA (2012b) The electro-oxidation of isotopically labeled glycerol on platinum: New information on C–C bond cleavage and CO2 production. Electrochem Commun 15:14–17. https://doi.org/10.1016/j.elecom.2011.11.013

Fernández PS, Tereshchuk P, Angelucci CA et al (2016) How do random superficial defects influence the electro-oxidation of glycerol on Pt(111) surfaces? Phys Chem Chem Phys. https://doi.org/10.1039/C6CP04768H

Figueiredo MC, Sorsa O, Arán-Ais RM et al (2015) Trimetallic catalyst based on PtRu modified by irreversible adsorption of Sb for direct ethanol fuel cells. J Catal 329:69–77. https://doi.org/10.1016/j.jcat.2015.04.032

Fontes EH, da Silva SG, Spinacé EV, et al (2016) In Situ ATR-FTIR studies of ethanol electro-oxidation in alkaline medium on PtRh/C electrocatalyst prepared by an alcohol reduction process. Electrocatalysis 1–8. https://doi.org/10.1007/s12678-016-0308-z

Garcia AC, Birdja YY, Tremiliosi-Filho G, Koper MTM (2017) Glycerol electro-oxidation on bismuth-modified platinum single crystals. J Catal 346:117–124. https://doi.org/10.1016/j.jcat.2016.12.013

Garcia AC, Kolb MJ, van Nierop y Sanchez C, et al (2016) Strong impact of platinum surface structure on primary and secondary alcohol oxidation during electro-oxidation of glycerol. ACS Catal 6:4491–4500. https://doi.org/10.1021/acscatal.6b00709

Gomes JF, de Paula FBC, Gasparotto LHS, Tremiliosi-Filho G (2012) The influence of the Pt crystalline surface orientation on the glycerol electro-oxidation in acidic media. Electrochim Acta 76:88–93. https://doi.org/10.1016/j.electacta.2012.04.144

Gomes JF, Garcia AC, Gasparotto LHS et al (2014) Influence of silver on the glycerol electro-oxidation over AuAg/C catalysts in alkaline medium: a cyclic voltammetry and in situ FTIR spectroscopy study. Electrochim Acta 144:361–368. https://doi.org/10.1016/j.electacta.2014.08.035

Gomes JF, Martins CA, Giz MJ et al (2013) Insights into the adsorption and electro-oxidation of glycerol: self-inhibition and concentration effects. J Catal 301:154–161. https://doi.org/10.1016/j.jcat.2013.02.007

González-Cobos J, Baranton S, Coutanceau C (2016) Development of bismuth-modified PtPd nanocatalysts for the electrochemical reforming of polyols into hydrogen and value-added chemicals. Chem ElectroChem 3:1694–1704. https://doi.org/10.1002/celc.201600147

Goulet M-A, Kjeang E (2014) Co-laminar flow cells for electrochemical energy conversion. J Power Sources 260:186–196. https://doi.org/10.1016/j.jpowsour.2014.03.009
Gupta M, Kumar N (2012) Scope and opportunities of using glycerol as an energy source. Renew Sustain Energy Rev 16:4551–4556. https://doi.org/10.1016/j.rser.2012.04.001
Hamann CH, Hamnett A, Vielstich W (2007) Electrocatalysis and reaction mechanisms. In Electrochemistry, 2nd edn. Wiley, Weinheim
Holade Y, Morais C, Servat K et al (2013) Toward the electrochemical valorization of glycerol: fourier transform infrared spectroscopic and chromatographic studies. ACS Catal 3:2403–2411. https://doi.org/10.1021/cs400559d
Horn EJ, Rosen BR, Baran PS (2016) Synthetic organic electrochemistry: an enabling and innately sustainable method. ACS Cent Sci 2:302–308. https://doi.org/10.1021/acscentsci.6b00091
Huang L, Sun J-Y, Cao S-H et al (2016) Combined EC-NMR and in situ FTIR spectroscopic studies of glycerol electrooxidation on Pt/C, PtRu/C, and PtRh/C. ACS Catal 6:7686–7695. https://doi.org/10.1021/acscatal.6b02097
Hutchings GJ (2008) Nanocrystalline gold and gold palladium alloy catalysts for chemical synthesis. Chem Commun 1148–1164. https://doi.org/10.1039/b712305c
Ilie A, Simoes M, Baranton S et al (2011) Influence of operational parameters and of catalytic materials on electrical performance of direct glycerol solid alkaline membrane fuel cells. J Power Sour 196:4965–4971. https://doi.org/10.1016/j.jpowsour.2011.02.003
Independent Review Panel Summary Report (2009) Current (2009) State-of-the-art hydrogen production cost estimate using water electrolysis
Iwasita T, Nart FC (1997) In situ infrared spectroscopy at electrochemical interfaces. Prog Surf Sci 55:271–340. https://doi.org/10.1016/S0079-6816(97)00032-4
Jin M, Zhang H, Xie Z, Xia Y (2011) Palladium concave nanocubes with high-index facets and their enhanced catalytic properties. Angew Chem Int Ed 50:7850–7854. https://doi.org/10.1002/anie.201103002
Kishida H, Jin F, Zhou Z et al (2005) Conversion of glycerin into lactic acid by alkaline hydrothermal reaction. Chem Lett 34:1560–1561. https://doi.org/10.1246/cl.2005.1560
Kjeang E (2014) Microfluidic fuel cells and batteries, spring briefs
Kongjao S, Damronglerd S, Hunsom M (2011) Electrochemical reforming of an acidic aqueous glycerol solution on Pt electrodes. J Appl Electrochem 41:215–222. https://doi.org/10.1007/s10800-010-0226-3
Kwon Y, Hersbach TJP, Koper MTM (2014) Electro-oxidation of glycerol on platinum modified by adatoms: activity and selectivity effects. Top Catal 57:1272–1276. https://doi.org/10.1007/s11244-014-0292-6
Kwon Y, Koper MTM (2010) Combining voltammetry with HPLC: application to electro-oxidation of glycerol. Anal Chem 82:5420–5424. https://doi.org/10.1021/ac101058t
Lam CH, Bloomfield AJ, Anastas PT (2017) A switchable route to valuable commodity chemicals from glycerol via electrocatalytic oxidation with an earth abundant metal oxidation catalyst. Green Chem 19:1958–1968. https://doi.org/10.1039/C7GC00371D
Lertthahan P, Yongprapat S, Therdthianwong A, Therdthianwong S (2017) Pt-modified Au/C catalysts for direct glycerol electro-oxidation in an alkaline medium. Int J Hydrog Energy 42:9202–9209. https://doi.org/10.1016/j.ijhydene.2016.05.120
Lima da Silva A, Müller IL (2010) Operation of solid oxide fuel cells on glycerol fuel: a thermodynamic analysis using the Gibbs free energy minimization approach. J Power Sour 195:5637–5644. https://doi.org/10.1016/j.jpowsour.2010.03.066
Marchionni A, Bevilacqua M, Bianchini C et al (2013) Electrooxidation of ethylene glycol and glycerol on Pd–(Ni–Zn)/C anodes in direct alcohol fuel cells. Chemsuschem 6:518–528. https://doi.org/10.1002/cssc.201200866
Marshall AT, Haverkamp RG (2008) Production of hydrogen by the electrochemical reforming of glycerol–water solutions in a PEM electrolysis cell. Int J Hydrog Energy 33:4649–4654. https://doi.org/10.1016/j.ijhydene.2008.05.029

Martins CA, Fernández PS, de Lima F et al (2014a) Remarkable electrochemical stability of one-step synthesized Pd nanoparticles supported on graphene and multi-walled carbon nanotubes. Nano Energy 9:142–151. https://doi.org/10.1016/j.nanoen.2014.07.009

Martins CA, Fernández PS, Troiani HE et al (2014b) Ethanol versus glycerol: understanding the lack of correlation between the oxidation currents and the production of $CO_2$ on Pt nanoparticles. J Electroanal Chem 717–718:231–236. https://doi.org/10.1016/j.jelechem.2014.01.027

Martins CA, Giz MJ, Camara GA (2011) Generation of carbon dioxide from glycerol: evidences of massive production on polycrystalline platinum. Electrochim Acta 56:4549–4553. https://doi.org/10.1016/j.electacta.2011.02.076

Matsuoka K, Iriyama Y, Abe T et al (2005) Alkaline direct alcohol fuel cells using an anion exchange membrane. J Power Sources 150:27–31. https://doi.org/10.1016/j.jpowsour.2005.02.020

Maya-Cornejo J, Guerra-Balcázar M, Arjona N et al (2016) Electrooxidation of crude glycerol as waste from biodiesel in a nanofluidic fuel cell using Cu@Pd/C and Cu@Pt/C. Fuel 183:195–205. https://doi.org/10.1016/j.fuel.2016.06.075

Mello GAB, Fernández PS, Martins ME, Camara GA (2017) Glycerol electrooxidation on platinum-tin electrodeposited films: inducing changes in surface composition by cyclic voltammetry. Electrocatalysis 8:1–10. https://doi.org/10.1007/s12678-016-0332-z

Monzó J, Malewski Y, Vidal-Iglesias FJ et al (2015) Electrochemical oxidation of small organic molecules on Au nanoparticles with preferential surface orientation. Chem ElectroChem 2:958–962. https://doi.org/10.1002/celc.201500084

Oliveira M (2016) Revista Fapesp, Ed. 245

Park JY, Zhang Y, Grass M et al (2008) Tuning of catalytic CO oxidation by changing composition of Rh–Pt bimetallic nanoparticles. Nano Lett 8:673–677. https://doi.org/10.1021/nl073195i

Pilatowsky I, Romero RJ, Isaza CA, et al (2011) Cogeneration fuel cell-sorption air conditioning systems. Springer Science & Business Media

Qi J, Xin L, Zhang Z et al (2013) Surface dealloyed PtCo nanoparticles supported on carbon nanotube: facile synthesis and promising application for anion exchange membrane direct crude glycerol fuel cell. Green Chem 15:1133–1137. https://doi.org/10.1039/C3GC36955B

Roquet L, Belgsir EM, Léger J-M, Lamy C (1994a) Kinetics and mechanisms of the electrocatalytic oxidation of glycerol as investigated by chromatographic analysis of the reaction products: Potential and pH effects. Electrochim Acta 39:2387–2394. https://doi.org/10.1016/0013-4686(94)E0190-Y

Roquet L, Belgsir EM, Léger J-M, Lamy C (1994b) Kinetics and mechanisms of the electrocatalytic oxidation of glycerol as investigated by chromatographic analysis of the reaction products: Potential and pH effects. Electrochim Acta 39:2387–2394. https://doi.org/10.1016/0013-4686(94)E0190-Y

Sau TK, Rogach AL (2010) Nonspherical noble metal nanoparticles: colloid-chemical synthesis and morphology control. Adv Mater 22:1781–1804. https://doi.org/10.1002/adma.200901271

Shan X, Díez-Pérez I, Wang L et al (2012) Imaging the electrocatalytic activity of single nanoparticles. Nat Nanotechnol 7:668–672. https://doi.org/10.1038/nnano.2012.134

Simões M, Baranton S, Coutanceau C (2011) Enhancement of catalytic properties for glycerol electrooxidation on Pt and Pd nanoparticles induced by Bi surface modification. Appl Catal B Environ 110:40–49. https://doi.org/10.1016/j.apcatb.2011.08.020

Simões M, Baranton S, Coutanceau C (2012) Electrochemical valorisation of glycerol. Chemsuschem 5:2106–2124. https://doi.org/10.1002/cssc.201200335

Soares LA, Morais C, Napporn TW et al (2016) Beneficial effects of rhodium and tin oxide on carbon supported platinum catalysts for ethanol electrooxidation. J Power Sour 315:47–55. https://doi.org/10.1016/j.jpowsour.2016.03.013

Solla-Gullón J, Vidal-Iglesias FJ, Herrero E, Feliu JM, Aldaz A (2013) Polymer electrolyte fuel cells: science, applications, and challenges. In: Franco AA. Pan Stanford

Song C (2002) Fuel processing for low-temperature and high-temperature fuel cells. Catal Today 77:17–49. https://doi.org/10.1016/S0920-5861(02)00231-6

Thia L, Xie M, Liu Z et al (2016) Copper-modified gold nanoparticles as highly selective catalysts for glycerol electro-oxidation in alkaline solution. ChemCatChem 8:3272–3278. https://doi.org/10.1002/cctc.201600725

Vidal-Iglesias FJ, Solla-Gullón J, Herrero E et al (2010) Pd adatom decorated (100) preferentially oriented Pt nanoparticles for formic acid electrooxidation. Angew Chem Int Ed 49:6998–7001. https://doi.org/10.1002/anie.201002501

Wang Y, Chen KS, Mishler J et al (2011) A review of polymer electrolyte membrane fuel cells: technology, applications, and needs on fundamental research. Appl Energy 88:981–1007. https://doi.org/10.1016/j.apenergy.2010.09.030

Wang Z, Xin L, Zhao X et al (2014) Carbon supported Ag nanoparticles with different particle size as cathode catalysts for anion exchange membrane direct glycerol fuel cells. Renew Energy 62:556–562. https://doi.org/10.1016/j.renene.2013.08.005

Wang Y, Leung DYC, Xuan J, Wang H (2016) A review on unitized regenerative fuel cell technologies, part B: unitized regenerative alkaline fuel cell, solid oxide fuel cell, and microfluidic fuel cell. Renew Sustain Energy Rev. https://doi.org/10.1016/j.rser.2016.11.054

Won JY, Sohn HJ, Song RH, Woo SI (2009) Glycerol as a bioderived sustainable fuel for solid-oxide fuel cells with internal reforming. Chemsuschem 2:1028–1031. https://doi.org/10.1002/cssc.200900170

Xin L, Zhang Z, Wang Z, Li W (2012) Simultaneous generation of mesoxalic acid and electricity from glycerol on a gold anode catalyst in anion-exchange membrane fuel cells. ChemCatChem 4:1105–1114. https://doi.org/10.1002/cctc.201200017

Yu Y, Gao Y, Hu K et al (2015) Electrochemistry and electrocatalysis at single gold nanoparticles attached to carbon nanoelectrodes. ChemElectroChem 2:58–63. https://doi.org/10.1002/celc.201402312

Yuksel A, Koga H, Sasaki M, Goto M (2010) Hydrothermal electrolysis of glycerol using a continuous flow reactor. Ind Eng Chem Res 49:1520–1525. https://doi.org/10.1021/ie9016418

Zalineeva A, Baranton S, Coutanceau C (2013) Bi-modified palladium nanocubes for glycerol electrooxidation. Electrochem Commun 34:335–338. https://doi.org/10.1016/j.elecom.2013.07.022

Zanata CR, Fernández PS, Troiani HE et al (2016) Rh-decorated PtIrOx nanoparticles for glycerol electrooxidation: Searching for a stable and active catalyst. Appl Catal B Environ 181:445–455. https://doi.org/10.1016/j.apcatb.2015.08.021

Zhang J, Liang Y, Li N et al (2012a) A remarkable activity of glycerol electrooxidation on gold in alkaline medium. Electrochim Acta 59:156–159. https://doi.org/10.1016/j.electacta.2011.10.048

Zhang Z, Xin L, Li W (2012b) Supported gold nanoparticles as anode catalyst for anion-exchange membrane-direct glycerol fuel cell (AEM-DGFC). Int J Hydrog Energy 37:9393–9401. https://doi.org/10.1016/j.ijhydene.2012.03.019

Zhang Z, Xin L, Li W (2012c) Electrocatalytic oxidation of glycerol on Pt/C in anion-exchange membrane fuel cell: cogeneration of electricity and valuable chemicals. Appl Catal B Environ 119:40–48. https://doi.org/10.1016/j.apcatb.2012.02.009

MIX
Papier aus verantwortungsvollen Quellen
Paper from responsible sources
FSC® C105338

If you have any concerns about our products,
you can contact us on
**ProductSafety@springernature.com**

In case Publisher is established outside the EU,
the EU authorized representative is:
**Springer Nature Customer Service Center GmbH**
**Europaplatz 3, 69115 Heidelberg, Germany**

Printed by Libri Plureos GmbH
in Hamburg, Germany